Safety in the Artroom

Safety in the Artroom

REVISED EDITION

Charles A. Qualley

Davis Publications, Inc.   Worcester, Massachusetts

This book was written to provide the most current and accurate information available about health and safety hazards in art-making spaces and other places where groups of children, young people, and adults of all ages may make art. However, the author and publisher can take no responsibility for any harm or damage that might be caused by the use or misuse of any information contained herein.

It is not the purpose of this book to provide medical diagnosis or information, or to set health or safety standards for any art-making space. Readers should seek advice from physicians, safety professionals, or environmental health specialists concerning specific problems and should in all cases carefully read and follow product information and instructions describing safe-use practices. Questions about specific products should be directed to the manufacturer. Nor is any of the legal information provided here intended to do more than point to information teachers, students, and parents might pursue with their own attorneys to redress problems they encounter.

Publisher: Wyatt Wade
Managing Editor: David Coen
Production & Manufacturing Manager: Georgiana Rock
Production Assistance: Books By Design, Inc.
Editorial Assistance: Annette Cinelli
Design: Books By Design, Inc.

Library of Congress Control Number: 2005932808
ISBN 0-87192-718-7
10 9 8 7 6 5 4 3 2 1
Printed in the United States of America

In memory of my late wife, Betty, whose support and encouragement are so deeply missed, and for my family Jan and Shane, John and Toni.

Contents

Part Three

Issues of the Health and Safety Content of the Curriculum and the Legal Problems Teachers and Leaders May Face 101

Acknowledgments for the Revised Edition

The persons who are mentioned in the acknowledgments in the first edition were instrumental in bringing this book to life. Though it is true that I no longer have close association with many of them, without their help, the first edition could not have led to this second edition. I am particularly indebted to my publisher, Wyatt Wade, president of Davis Publications, who has encouraged me to undertake this task for some time, and to Carol Traynor and David Coen, my editors, as well as all those involved in its production and design. Without their support, this updated and expanded edition, reorganized and presented in a manner that will make it easier to use and of value to a broader group of people making and teaching art, would never have been undertaken.

Thank you to all who have been my teachers and mentors; prominent among these are all of the students I have had throughout my career in art education and my many colleagues and friends.

Charles A. Qualley
Albuquerque, NM
2005

Acknowledgments for the Original Edition

Ray Hall and David Baker are nearly as responsible as I am for this book. I am deeply grateful to have them as friends and mentors.

As early as 1969, Ray Hall, then Safety Officer at the University of Colorado, urged me to write a book on the health and safety hazards involved in art-making processes. Ray was among the first to be aware of those dangers and felt that information should be available to everyone working in school environments.

I first proposed to David Baker, editor of *School Arts,* that I write a limited series of health and safety-related articles for the magazine in 1976. His enthusiastic acceptance of that idea led to the "Safetypoint" series, which have been used extensively in writing this book. Dave has been supportive and encouraging, as have all the people at Davis Publications.

Others, too, have made important contributions to this book and if I do not mention all of them, it is only because there are so many. My wife, Betty, has been my constant teacher and critic in the writing of this book. Michael McCann, who reviewed the manuscript, and Monona Rossol have added immeasurably to my understanding of art materials and processes, which are inherently hazardous. And over the years many of my students have shared with me their experiences and their attitudes about hazards in the artroom.

Marilyn Newton, Carol Worlock, and Karen Greene gave me a number of very helpful insights into how art teachers deal every day with hazards in their artrooms. Marc Swadener worked with me to devise procedures for the analysis of research data.

Comments from readers of my *School Arts* articles have also been very important to me.

Foreword to the Revised Edition

Over the last three decades artists and art teachers have come to recognize that the environment in which art is created is not without some hazard to themselves and their students. In fact, we understand better than ever before that the materials we use, the ways we use them, the very places we work, and our interaction with others in the work space are often sources of immediate and long-term health problems.

A quick Internet search under the keywords *art materials safety* will produce over 800,000 hits; surely this is an indication that materials we work with can be hazardous. When the first edition of this book was published, that information was not available to those of us making art or teaching it. It was difficult to find out which materials or processes could cause teachers and students trouble. Many did not believe there were problems; some refused to take any of these concerns seriously and some ridiculed health and safety concerns as nonsense and an intrusion on their creativity or that of the students they taught. Some of those critics are no longer here. Some carry in their bodies the damage that their lack of understanding resulted in. We hope their students do not still face health problems that could have been avoided.

We frequently hear about newly identified carcinogens and other dangerous substances in our environment. Accidents regularly claim lives or limbs in homes, schools, or workplaces. Few of us are particularly shocked any longer. There seems to be nothing we do that is not in some way harmful, and many feel there is little we can do about it.

In the foreword to the first edition of this book, I wrote: "This book was intended to help art teachers understand the hazards that exist in art processes. It suggests ways to eliminate or deal with them safely. Few claim the artroom can be made completely hazard-free, since we know too well how inventive all of us are in discovering ways to injure ourselves. Nonetheless, teachers can establish procedures in the artroom by which students learn safe and effective ways of doing art activities. Most hazards are unnecessary, and in those processes where hazards are inherent, it is usually possible to minimize them." This remains true today and for this edition, but the focus has been expanded to include additional art-making spaces in which teachers and students work and to provide additional information and resources that were not available when the first edition was published.

Artists and art teachers use chemicals every day in their art-making spaces. The hazards of some of them have become familiar and others cause questions and concern. We know of both teachers and students who have had adverse reactions to certain materials. Some have suffered serious illness most likely caused by substances used in making art. Some have died. This new worry augments our old concern about injuries to our students.

As art teachers in whatever setting we work, we have a responsibility to do all we can to spare students and ourselves illness or injury. Consequently, we want to be sure everyone learns safe ways of making art. Remember, it is the artist and the art teacher who are the most exposed and most at risk; we must worry about the students, but we cannot forget ourselves.

How to Access Reference Sources

Throughout this book you will find in-text references that refer to sources found in the printed bibliography. On many pages you will also find footnotes, which will provide information about sources of more extensive information than the book contains, as well as printable forms that can be used to enhance your health and safety program.

These footnoted sources are available through online links that are found in the designated location on the Davis Publications Web site (www.davisart.com/safety). These links will be updated regularly to provide readers the opportunity to explore the referenced time-sensitive material much more fully than is possible in the book itself.

In some cases, the referenced information will relate to commercial products and commercial Web sites. These have been included only for the purpose of providing examples of types of equipment that are mentioned in the book and do not in any way represent endorsements of those products.

Foreword to the Original Edition

Our environment is not particularly benign. Every day newspapers report new potential carcinogens and other dangerous substances in our environment. Accidents regularly claim lives or limbs in homes, schools, or workplaces. Few of us are particularly shocked any longer. There seems to be nothing we do that is not in some way harmful. There sometimes seems still less we can do about it.

This book is intended to help art teachers understand the hazards that exist in art processes. It suggests ways to eliminate or deal with them safely. Few claim the artroom can be made completely hazard-free, since we know too well how inventive all of us are in discovering ways to injure ourselves. Nonetheless, teachers can establish procedures in the artroom by which students learn safe and effective ways of doing art activities. Most hazards are unnecessary, and in those processes where hazards are inherent, it is usually possible to minimize them.

Art programs need not be destroyed in the elimination of hazardous art materials or practices. Correcting problems is not always expensive, nor should calling attention to them necessarily threaten an art teacher's program or job. The chapters ahead will identify potential problems and point out solutions. Such an approach will not only reduce the possibility of illness or accidents but may even result in a better overall art program.

Some 100,000 man-made chemicals exist. Some are unquestionably beneficial—we use them to heal the sick, to produce fabrics, and to grow food. Many of these chemicals, though, are dangerous and are of real concern to us.

As artists and teachers, we use chemicals every day in the classroom. The hazards of some of them have become a worry. We know of both teachers and students who have had adverse reactions to certain materials. Some have suffered serious illness probably caused by substances used in making art. This new worry augments our old concern about injuries. As art teachers, we wonder if we are doing all we can to spare students and ourselves illness or injury. Consequently, we want to be sure everyone learns safe ways of working in the artroom.

Art should be a joyful experience for both student and teacher. It provides ways for students to experience a richer world. Injury or illness should not be allowed to spoil this experience.

Part One

An Overview of Health and Safety Issues in Art-Making Spaces

Chapter 1

Identifying the Hazards

Although the major focus of this book is on art programs during the school years, k–12, remember that art is made by individuals and groups, young people as well as adults, and in a variety of different settings. In recent years the demographics of art making have changed and so, too, has the need to help all art teachers and art activity leaders become aware of all potential hazards.

Children in day care centers who may be anywhere from two to five years old are painting, drawing, cutting paper, making assemblages and collages from a variety of materials, and making clay objects and sometimes even glazing them. Children in this age group are at high risk of illness and injury for a variety of reasons, and their teachers are often poorly informed about the potential health and safety issues with which they should be familiar.

After-school meetings of Boy and Girl Scouts, Campfire Girls, and their various separate age-based sub-organizations, day camps, and similar organized activities for children focus a great deal of their time on doing art-like projects. While many of these projects are not art education in a strict sense, they do involve the use of art materials that carry the same hazards as do those used in school. Art activity leaders need to know what they are asking their children to do and how the materials used may put them in harm's way.

Summer children and youth programs, such as those in churches or in camps, also often have art-making projects at the center of their activities. Leaders of these programs, too, must be guided by a familiarity with the

problems of the materials and processes they are using. Because these programs often are held outside, leaders sometimes assume that the materials
used are less hazardous, and that is, to some extent, true. However, particularly in these temporary outside facilities, the potential for safety problems may actually be increased. A thorough knowledge of safe practices,
as well as hazards produced by materials, is a necessity for each group
leader.

Adult classes are usually organized and held in community college or
high school artrooms or similar facilities in the evenings or on weekends
with professional instructors. Certainly these instructors should be aware
of the potential hazards involved in this work and plan, as they do for
their normal everyday classes, to keep adult students informed about the
materials and processes they are using. Because adults are at a different
level of risk than are younger students, instructors must be knowledgeable
about those differences. Sometimes art classes, especially pottery and
painting, are held in storefront studios where adults drop in and work on
their own objects. Sometimes the students simply apply glaze to a pre-
formed one, but in all of these places art activity leaders still need to know
the precautions that must be taken. These spaces are often poorly
equipped and lighted, ventilation is inadequate, and fire danger is high.
This places a special responsibility on the art activity leader to ensure that
the necessary precautions are taken.

Senior citizen centers are almost always the home of one or more art
activity programs including pottery of all types, drawing, painting, craft
making, and the like. In the same way that children are at high risk of
injury or sickness brought on by certain art materials, older adults are also
at high risk because of preexisting illness or infirmities and the common
physical characteristics of this age group. It is essential that special care be
taken not to exacerbate those problems.

Individual homes are the site of many art activities, especially if the parents are interested in encouraging their children to develop their skills and
talents beyond what they can receive in schools. While many of the concerns about art making in a group setting such as school come from the
interaction within the group itself, not all of them do. Even working alone,
children must follow safe practice guidelines and understand the correct
way to use art materials and tools. Parents seeking to encourage their children's development must be knowledgeable about safety precautions to
avoid unwittingly causing damage to their children's health.

The art-making space, wherever it is located and depending on the types of activities that are undertaken, often more closely resembles an industrial environment than it does regular classroom space. If the space is in a church basement, a senior center, a storefront room, or an individual home, it is a unique space with unique demands. Many people, sometimes including school principals or persons responsible for other kinds of spaces, are unaware that making art may require more than an easel, a single brush and pan of watercolor, or a box of crayons. Of all the different educational settings in a school, for example, the artroom, along with the industrial arts shop and the chemistry lab, relies on specialized equipment, tools, and materials. Non-school art-making spaces have the same special requirements. With all student age groups, the art-making space should always be understood to be different from the regular classroom. Whatever the intensity or seriousness of the art activities being undertaken, precautions must be taken to ensure that the working space and materials and equipment used do not lead to student illness or injury.

Some art activities are obviously more complex than others and require many more tools and materials. As complexity increases, so do the hazards. If all art making were drawing with a #2 pencil or painting with watercolor, there would be few problems to concern teachers or art activity leaders. But art content has broadened to include many materials and techniques once thought limited to industrial operations, and all of them place different and serious obligations on the teachers. Most non-school art activities are unlikely to be in this category, but some, such as those done in adult and older adult classes, may be. There are art teachers and art activity leaders who are aware that health and safety risks exist in their art work space but who choose to do nothing about them. They fear the cost of making corrections will either force the abandonment of parts of their program or art activity offerings or give an unsympathetic administrator the chance to eliminate art altogether. That fear may be well founded in some situations, but in most cases it is not. In any event, a teacher or art activity leader who intentionally does nothing invites serious consequences; some risks may well result in student illness or injury and ultimately in the unpleasantness of litigation. (See Chapter 9 for a discussion of teacher liability concerns.) If there is a risk that a program might be cut when requests have been made to correct hazards, the program almost certainly will be cut if an administrator finds that teachers are knowingly perpetuating unsafe conditions or practices.

Art activity leaders and art teachers who are uninformed about potential hazards jeopardize both themselves and their students. Each teacher has the responsibility to know his or her own program well enough to understand the various hazards and deal effectively with them. With good planning and careful methods of working, the art program and group art activities should not suffer for making the necessary corrections, and the students will benefit substantially from the entire process. Moreover, since the teacher is in contact with the hazards in the artroom for the longest period of time, the payoff for being careful and aware may actually be a healthier and longer life for her or him.

Defining and Clarifying the Terms *Health* and *Safety*

It is a little artificial to separate the health problems and the safety hazards in the art-making space. Each can result in injury, and often actions taken to eliminate one, eliminate the other. For example, removal of sawdust to reduce the safety hazard of falling on slippery floors also reduces the health hazard caused by breathing dust particles. However, it is customary to distinguish between health and safety hazards by the effects each has on the human body, and that is the distinction that is made here: hazards resulting in illness are considered health problems, while those causing injury are considered safety problems.

Understanding the Role of Federal and State Regulations

Three federal agencies are charged with promulgating and enforcing health and safety regulations that affect nearly all the citizens of the United States. OSHA (Occupational Safety and Health Administration) is an agency of the U.S. Department of Labor and NIOSH (National Institute for Occupational Safety and Health) is a part of the Centers for Disease Control and Prevention, and along with the EPA (Environmental Protection Agency), fulfill these functions nationally. In addition, in many states there are agencies that enforce their state's regulations or, in partnership with the federal agencies, enforce federal regulations that apply to health and safety issues within their jurisdictions.

In nearly all cases, these agencies do not have any jurisdiction over the health and safety of students or teachers in schools. Their charge is to oversee adults in the private and sometimes public workplace and they do not deal with schoolroom health and safety issues. Some state departments of education, the state agency charged with developing rules and policies for schools, have health and safety plans, but none of these are more than generally applicable to the artroom. Even where there may be some applicability to artrooms in public schools, these plans do not apply to art that is taught in places other than the schools.

This does not mean that art teachers and art group leaders should not be aware of the work of these federal and state agencies nor of the information they can provide that will help in creating a healthy and safe place for art to be made. Although little of their information applies directly to art-making processes, some of it is extremely useful when extrapolated to art-making settings. NIOSH, for example, has provided a series of checklists for schools, some of which are directly applicable to activities in the art-making space,* and the EPA has information that is useful in assessing concerns about ventilation.† Art teachers and art group leaders should become familiar with this information in planning art activities.

The Responsibility of Program Administrators

The legal responsibility of program administrators is unclear, except as it may be governed by those legal concerns discussed in Chapter 9. However, there is no question that each administrator has an obligation to the students and teachers in his or her program to make sure that the space is as safe as possible. Whether it involves health concerns or safety issues, the administrator is the person who makes the decisions that will fund or not fund corrections to poor conditions in the art-making space. To do this, all administrators must be familiar with the federal and state regulations that even marginally apply to their situation and they must seek advice from experts about any problems they have identified. Failing to do this leaves them open not only to legal liability issues, but to negative moral judgment as well. Any administrator ignoring either the suggestions or guidance of federal or state regulations designed to protect the general public does not demonstrate the understanding of responsibility that goes with his or her administrative position and has no justification in holding it. It is the responsibility of art teachers and art group leaders to make sure their program administrator is aware of his or her responsibility and has the information necessary to correct any problems that exist.

---

* NIOSH Checklist Program for the Schools, Table of Contents (www.davisart.com/safety)
† EPA Ventilation Checklist and Log (www.davisart.com/safety)

The Need for Action by Teachers, Parents, and Art Activity Leaders

Even with this brief discussion of the variety of places art is made and the general comments about how art-making programs differ or are similar to other student activities, it must be confusing to those with little formal training in art when they are confronted with art hazards and what to do about them. The following chapters provide information that will give teachers and art activity leaders the confidence to work in comfort with students of all ages. The knowledge that the art materials and processes have been carefully planned to reduce or eliminate illness or injury to the students will give teachers the confidence to spend more time helping students with the art-making processes and techniques for undertaking meaningful art activities.

No thinking adult would intentionally ignore safety and health problems for his or her students; nonetheless these issues are often overlooked. This is why teachers and leaders involved in art programs, wherever they may take place, must understand and use the information here to guide their teaching. When they do become knowledgeable about these safe practices, they should insist that others teaching around or with them follow the same guidelines. Never should illness or injury be allowed to diminish the joy all people derive from their art-making experiences.

Chapter 2

Precautions and Protection

The best way to eliminate a hazard is to eliminate its cause. Evaluate any hazardous activity or material in use and determine if its educational or experiential value is worth the time, money, and effort necessary to overcome the problem it creates. The philosophy of this book is to emphasize several fairly simple methods for reducing art work space health and safety hazards. These suggestions are based on good sense, good housekeeping, and the importance of eliminating as many activities involving hazards as is consistent with good art education and good art-making experiences. Keeping the art-making space clean and orderly is one of the most effective means of controlling hazards. Teaching all students to be responsible in their handling and use of materials and tools is yet another important aspect of safe art making.

Few students respond well to nagging, but reminders must be constantly visible. Signs are a useful device for this purpose. And because accidents sometimes do occur, there must be a plan to deal with injury. The teacher should also understand basic first aid concepts that apply to the art-making space. These various elements are all important in eliminating hazards and must be a part of a health and safety program.

Improvements to Health and Safety Conditions Without Cost

Before a good case can be made for spending any funds to reduce hazards, teachers must achieve all that can be done without cost. Clean the art-making space; put supply cabinets in order and label all supplies; reorganize activities to limit or confine problems; and get rid of unused or unusable materials. An objective appraisal will identify much that is cluttering the work space, creating dust and difficult working conditions.

When these first steps have been taken, it will be easier to enlist the cooperation of others. The inertia will have been broken. Few teachers enjoy art space housekeeping, but a positive attitude toward it is necessary if health and safety hazards are to be controlled.

Clay, plaster, fiber, and everyday dust are usually serious art work space problems; reorganize work procedures so that unavoidable dust is not unnecessarily stirred up. Confine activities like clay mixing to a separate room limited to times when few students are around. These actions cost no money but make obvious and important contributions to hazards elimination. Other points:

• Is the disposal of waste material done properly?

• Are tools properly stored?

• Are there room and materials inventories?

• Have you developed a set of records showing the scope of your health and safety actions?

These no-cost improvements will indicate to school or program administrators a serious intent to correct hazards. It is the most effective step toward getting financial support to solve the remaining problems.

Signs

Signs are no substitute for a teacher's instructional responsibilities or student knowledge. But the selective use, proper placement, and periodic changing of signs can effectively remind students of hazards without resorting to nagging.

Any sign is only as effective as its placement and the clarity and pertinence of its message. Obviously, a sign must be displayed where it can be seen and the message must be quickly understood. Signs based on symbols rather than words may sometimes be a requirement. For example, some instructors will need to consider what kind of signs will be effective with pre-school children who do not yet read or with adult students who may not read English.

Signs must also be selective. This means the teacher or art activity leader must analyze the routine activities of the art program and establish a priority for signs. What are the signs for: to warn, to instruct, or to remind? A sign displayed on or near the paper cutter may be used either to warn of its hazards or to remind the student how to use it. "Danger: Watch your fingers!" might be alternated every three or four weeks with "Sharp blade. Keep fingers well back." Changing signs and their location reduces the problem of familiarity resulting in the message being overlooked. Signs that never change are often no longer seen and have little or no impact on student actions. They become so dust covered or so familiar a part of the room that no one ever notices them.

What sort of signs should be used? How many should there be? How often should they be changed? Each art-making space is different, so each teacher must decide what will work best. Some suggestions are found in Figure 1.

Clothing and Dress Restrictions

Many schools have dress codes. Whatever one's attitude is toward these regulations, the art teacher should establish a code applying specifically to clothing worn in the art-making space. Even in non-school settings such as camp or summer programs or adult classes, participants should have a clear understanding of appropriate dress. Such a code has far less to do with style than safety, nothing to do with propriety, but much to do with protection. These regulations should also apply to hair and jewelry, which also represent potential problems in working with tools and equipment.

Students may often not be aware of what can happen when long hair becomes entangled on the spindle of the bench grinder or a loose sleeve or a dangling bracelet is wrapped around the bit of a drill press. No one enjoys horror stories about students being scalped or mangled, but such things have happened.

Simple and well-enforced rules about clothing, jewelry, and hair are appropriate for the artroom. It is, after all, a work area and not a center for fashion display. To establish an artroom code, work with the students to identify hazards and develop rules to overcome them. Involving the students in this process usually makes rules more palatable to them. There don't need to have been injuries in the past to justify preventive measures; *the purpose of these rules is always to prevent accidents, not prove they can happen.* With hazards identified and a code established, a meeting with the principal or program administrator can result in designing enforcement methods and clarifying the role of the administration in that enforcement.

USING SIGNS IN THE ARTROOM

Sign Message	Possible Symbol	Suggested Placement	Frequency of Change
Eye protection; safety glasses. (WEAR YOUR SAFETY GLASSES)		Above benches in the area where they are to be used.	Rotate position every month or six weeks.
Fire extinguisher location. (FIRE EXTINGUISHER IS NEAR THE DOOR)		Near doors or flammable materials; close to extinguisher.	Change design twice a year.
Proper handling and storage of flammable materials. (KEEP CONTAINERS COVERED)		Inside the storage area used for these materials.	Change signs two or three times a year.
First aid equipment location. (FIRST AID STATION)		Close to first aid supplies.	At the beginning of each semester.
Exhaust hood fan. (TURN ON FAN)		On hood itself; at eye level.	Change design each semester.
Handling restrictions on equipment. (USE ONLY AFTER INSTRUCTION)	Fire Extinguisher, Remove pin and squeeze trigger.	At equipment site; on large equipment.	Each semester.

Note: The information and sample signs in this table are suggestions for what might be used in an art-making space. One way to call attention to the importance of signs is to institute a design project in which students design appropriate symbols for the signs in their own art-making space.

Figure 1

Once established, the code must be taken seriously. If hair must be tied back, see that it is; if loose clothing is not allowed, require the student to change or leave the working area; if bracelets or necklaces must be removed, be sure they are and provide a secure place to store them. There should be no exceptions, including exceptions for the art teacher or activity leader.

Accidents can occur in any setting, but there is never a reason for hair, loose clothing, or dangling jewelry to cause them. A well-thought-out dress code for the art-making work space should help ensure that they don't.

Protective Equipment

The need for individual protective equipment such as safety glasses, respirators, or gloves should never exist in any art program involving preschool or elementary-age children because materials requiring them should not be used at those levels. At the middle school level, some personal protective equipment, such as safety goggles and dust mask, will be necessary when working with ceramics, wood, or jewelry. But it is at the high school level or in art classes geared toward adults where many activities produce dust, flying chips, or fumes or require contact with hot or sharp materials. Serious consideration must be given to equipment and materials used in these types of activities.

Personal protective equipment shields the individual from direct exposure to unavoidable dust, mist, vapors, flying particles, chemicals, and noise. Generally, much of this equipment should be considered a "last resort" and used only when all other measures have proven inadequate. Good housekeeping, adequate ventilation, and substitution of materials must first be fully implemented, because they are fundamentally better techniques. Most personal protective devices are clumsy, uncomfortable, and expensive. Often they create problems, because students do not want to use them regularly, do not adapt to them well, and either actively resist using them or simply "forget." Sometimes using such equipment cannot be avoided, but it should never be considered a panacea for artroom hazards or a quick fix for environmental problems.

Eye protection is necessary in grinding and polishing jewelry, chipping and carving sculptured forms, and cutting and sanding wood. OSHA standards, in fact, require wearing safety goggles in most industrial settings, and students should do so in situations where the need is obvious and this form of protective equipment has become fairly standard. Only two points about safety goggles need to be made: be sure they are actually used and be sure to select the correct goggles for the job. They should be flexible fitting, have regular ventilation built in to prevent fogging, and have a wide angle of vision. However, the so-called eye-cup type goggle may be needed for certain welding and chipping activities because they fit closely to the face. Be sure the goggles fit properly, and give special attention to students who wear eyeglasses or contact lenses. Where chemical fumes are involved, students wearing contact lenses should not wear ventilated goggles, because vapors or gases can become concentrated under or absorbed by the lenses and cause serious eye irritation. Safety goggles are easily scratched and vision can be significantly impaired, so they must be replaced when this occurs. Elastic headbands eventually stretch and become loose, so they should frequently be replaced as well.

Respirators and masks filter out harmful vapors and dust, but the correct NIOSH-approved filter must be used or the expected protection may be missing. (NIOSH refers to the National Institute for Occupational Safety and Health.)* The fit is critical because a mask or respirator that allows dust, vapors, or other airborne material to enter around the edges has no value, and the user is not receiving the protection expected. Read product information carefully before ordering to be certain the filter selected will remove the type of dust or vapors that represent the hazard. These will probably be materials such as the free silica in clay powder or organic vapors from solvents and lacquer thinners. Do not expect that masks will filter out gases and fumes from kiln firing since too many different fumes are produced for any one mask to be effective, and masks can give a false sense of security.

Respirator use must be carefully limited because most of them make breathing difficult. Students with any respiratory or heart problems should use them only with their doctor's approval. Perhaps more than any other personal protective equipment, respirators should be used only under extreme situations: when local exhaust ventilation is impractical for some very limited or infrequently done operation or while a regular ventilation system is temporarily out of order. Although beards are seldom found on students in k–12 classrooms, there may be teachers in those classrooms and teachers and students in adult classes and senior center activities who may wear them, and teachers should be aware that a beard prevents a

---

* NIOSH Respirator Standards Development and Disposable NIOSH Respirator Filters and Dust Masks (www.davisart.com/safety)

proper fit around a respirator and completely negates any protection it might be thought to offer.

Gloves are available in a variety of materials, each designed to protect against specific chemical damage. In typical artroom activities, disposable vinyl gloves are usually adequate. These are thin enough to provide the tactile sensitivity necessary for most activities, are resistant to limited exposures of most chemicals, and are relatively inexpensive. For the limited amount of solvents used in most artrooms, these will provide the necessary protection if they are replaced as soon as they give any evidence of tearing or leakage. Be sure they are long enough to cover the wrists and the wearer is not allergic to vinyl.

Protective skin creams, or "barrier creams," are also available in cases where gloves cannot be used.* To be effective, they must be re-applied frequently and after any hand washing. For both gloves and barrier creams, know the specific chemicals being handled, and check manufacturer's information concerning the best glove material to use for the expected protection. Some gloves are very expensive, so be sure they will do the job.†

Noise from machine use will not be frequent or consistent enough to require ear protection for most art-making activities, but noise problems should not be overlooked. Grinders, buffers, saws, and sanders may require students who are using them to wear earplugs or sound-insulated ear covers. First, reduce machine noise through proper mounting and maintenance. Install sound-absorbing materials around the machine area, and limit the time any individual or class is exposed to the noise. Ear protectors should be available near any noisy machine so students can use them when they feel a need. But in most art-making spaces, their use need not be mandatory and teachers should make the decision about when they should be used.

Personal protective equipment should certainly be used in some regular art activities: dust masks for the mixing of clay or powdered dyes and pigments; gloves for working with oil-based printing inks and any solvents; and safety glasses for grinding or chipping various materials. Any other protective equipment should be used only when there is no other option. When it must be used, the fit must be perfect. Select this equipment carefully to protect against specific hazards, or it can be a waste of time and money.

---

* The Effectiveness of Barrier Creams (www.davisart.com/safety)
† York University Safety Notice: Gloves (www.davisart.com/safety)

Using a Personal Protective Equipment Checklist

One of the several NIOSH-developed checklists is available to use in evaluating the use and appropriateness of personal protective equipment.* The checklist provides guidance for equipment used by adults in a working environment, but is applicable to some situations in art-making spaces where sculpture and woodworking are done.

Teacher Training in the Use of Personal Protective Equipment

It is not enough to know that protective clothing or equipment should be worn in some circumstances. Neither is it enough to know what the correct equipment is and have it available when it is needed. Teachers and group leaders must also know how to use this equipment correctly, and they have an obligation to show students how to use it when they are participating in processes that require it.

Because it may be difficult to find specific training programs, teachers and group leaders need to read all instructional materials that come with the equipment. They should also contact others within the school district to assist them in learning to use the equipment properly. Teachers in science and industrial shop courses may be good sources of information and should be sought out for help. Failing this, it might be useful to contact the OSHA office for assistance. NIOSH posts some information on personal protective equipment that can be extrapolated to art-making space needs. If no one can be found who is able to give the teacher instructions for the proper use of the equipment, and information from other sources is not applicable, do not use the equipment and do not have students working in processes that require it.

Ventilation

Ventilation is the key to solving many of the most serious health hazards in the artroom. The warning found on labels to "use with adequate ventilation" calls attention to the problem but provides no understanding of what is adequate. Ventilation requirements differ according to the setting: size of the room, numbers and ages of the students, and the concentration of hazardous fumes. For this reason no specific standards for ventilation

---

* NIOSH Personal Protective Equipment Self Inspections; OSHA Personal Protective Equipment Sample Written Program (www.davisart.com/safety)

requirements exist and there are no federal laws to protect students from exposure to contaminants that pose potential health risks.

The EPA (Environmental Protection Agency) has prepared a checklist for schools to inspect and evaluate their ventilation system. This information is technical, but useful in understanding what must be considered in maintaining or improving the general ventilation in the school building.*

At one time, it may have been possible to say that with effective and complete ventilation, there is almost no art material that cannot be used safely when the right conditions prevail (Carnow, 1981), but it is difficult to know if that continues to be true because of new processes and materials. Furthermore, no art-making space will automatically have perfect ventilation for every substance and with some substances proper ventilation may not be easily achieved, even if there is good overall ventilation in the building. Teachers and leaders must know the contaminants that must be removed and what systems are necessary to do the job. Simply opening a window will seldom be sufficient, even when a fan is used.

In making plans for proper ventilation, start with the information gathered from an inventory of the materials used in the space. Try several sources for ventilation advice: the science department in one of your schools or the local department of public or environmental health, for example. In some states, the state board of education will have policies in place that may provide helpful guidance. But it is not easy or inexpensive to find people with expertise in designing art-making space ventilation systems. Even the most highly qualified industrial ventilation specialists will need specific information about the kind and amount of contaminants before they can make judgments and recommendations.

Ventilation: A Practical Guide (Clark, Cutter, and McGrane, 1984, pp. 21–24, 35, and 53), which is written for application to artists' studios, comes closest to addressing the needs of the art-making space. This book is no longer in print, but may be available in libraries, and the most useful sections have been referenced here. The ventilation requirements for many materials are complex, and the best and most practical approach is to eliminate problem materials.

When as many problem materials have been eliminated as is possible, determine if the general (room-wide) ventilation is sufficient or whether the activities that require materials producing fumes or dust can be concentrated in a smaller space. If they can be confined, a local exhaust system, such as a vent with fan, may be able to remove the contaminants before

* EPA Tools for Schools: Ventilation Checklist and Log (www.davisart.com/safety)

they disperse throughout the entire room. It is far easier to ventilate a small space than a large one. For example, the fumes from melting wax for batik should be drawn completely away from the space where people are working and should not be allowed to spread throughout the art-making space. Several ventilation methods are shown in Figure 2.

Consider electric kilns, which are so common in schools and in art-making spaces such as studios where private instruction is given or in senior center craft rooms: many are located inside the art work space so that a canopy hood or down draft venting system is necessary to remove the fumes generated. Kiln fumes present a special problem because they are released by a variety of substances in the firing process, and their exact nature is not often known. It might be possible to fire a kiln without a ventilating hood if it were done at night when the room and building are empty, but unattended firing is extremely hazardous and may well result in serious damage to the kiln or to the building itself.* As long as people remain anywhere in the building, they may well be in some danger from the fumes. It is also possible that fumes may linger and still be present in the building the next morning.

Before expensive solutions are undertaken, eliminate problem materials through substitution, and confine activities that generate fumes, gases, or dust to places where contamination is most easily extracted. Keep the room clean. Be guided in your planning by the following "Rules for Good Ventilation" (Clark, Cutter, and McGrane, 1984, pp. 22–23):

1. Direct air flow away from breathing zones of people who work in the area.
2. Exhaust contaminated air from the work space.
3. Place the exhaust opening of the ventilation system as close as possible to the source of the contaminants.
4. Avoid cross drafts.
5. Supply make-up air to replace the air exhausted by the ventilation system.
6. Discharge the contaminated air away from openings that draw air into the studio or shop.
7. Avoid polluting the community.

Ventilation systems produce heat loss in cold weather climates, because air replacing the exhausted air usually comes from the outside. How much does this amount to in added heating costs? To estimate this figure for a

* AMACO Kiln Venting information (www.davisart.com/safety)

POSSIBLE VENTILATION DESIGNS

General purpose ventilation with fan to outside.

Canopy hood for print inking, acid etching, ceramic kilns, foundries.

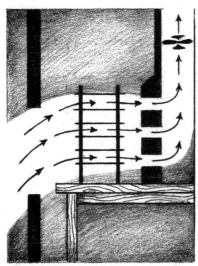

Slotted hood for drying prints, silkscreen printing, soldering.

Elephant trunk (moveable) exhaust for welding, woodworking machines.

Figure 2

specific location, a ventilation engineer can calculate hourly and yearly costs of tempering (heating up) the air by using equations designed for that purpose (Loeffler, 1984, pp. 7–11). This information is helpful in understanding the real cost of providing adequate ventilation to an art-making space.

To minimize this cost, system designers usually will try to use one of the following methods of conservation:

- Reduction in the volume of air handled. This may be done by reducing hours of use, using low-volume, high-velocity hoods, or using other principles of local exhaust capture.

- Delivery of untempered air. Using supply air that has not been pre-heated may be a way of reducing cost if that air is likely to undergo some heating in the room itself. If, for example, the ventilation system is running only when classes are in session, body heat may be sufficient to keep the room comfortably warm.

- Recovery of energy from exhausted air. Heat exchangers can be used to extract heat from outgoing air. This is usually prohibitive in an art-making space because of equipment cost.

- Recovery of uncontaminated air. Such a procedure requires primary and secondary air cleaning equipment with automatic monitoring to ensure the recirculated air is clean. The cost and complexity of such a system rules it out for the artroom application (Loeffler, 1984, pp. 7–15).

A combination of the first and second of these methods is the most applicable to artroom ventilation requirements. Some heat loss will occur, but that loss must be balanced against the health hazards involved. With careful operation of the ventilation system, energy loss can be kept to a minimum, and students can be both safe and comfortable.

Why You Need a Basic Knowledge of First Aid

In most schools, a nurse is usually on duty at least part of the day, which is not typically the case in other settings where art is made. Art teachers and activity leaders often develop a false sense of security with the knowledge that a nurse is in their building, and art activity leaders in other settings may tend to ignore or downplay the need for first aid response to injuries. When possible, of course, sending the student to a nurse is the wisest action, but sometimes this cannot be done. The nurse may not be in the building, a nurse may not be available, or the injury may be so severe that it requires instant attention. The teacher must know immediately what to do.

Long before the need arises, the art teacher should meet with a school nurse (or district nurse if the school does not have its own nurse), and art activity leaders should meet with a health care professional to get information and/or instruction on first aid. Contact the Red Cross for information on the availability of first aid classes. A first aid kit containing antiseptics, bandages, and compresses should be prepared for the room, and the teacher's or leader's instruction should include when and how to use them. Ideally, all art teachers or leaders in a community center should have a session together, during which they receive help and have the opportunity to practice techniques of immediate treatment. Apart from general first aid, teachers should be taught to respond to the specific types of injury or illness that may actually occur in the artroom: cuts, burns, shock, or the ingestion of toxic materials. Knowing CPR techniques may be extremely useful in everyday life, but it is far less likely to be needed in the art-making space than knowing how to deal with severe bleeding or a second degree burn.

Wounds and Bleeding

- Don't dismiss slight wounds and scratches. They should be washed thoroughly and covered with a sterile dressing. Encourage puncture wounds to bleed a little to flush out foreign material. Any wound has the potential for developing tetanus and should be treated accordingly.

- Apply a sterile dressing to prevent entrance of additional germs.

- Wash hands with soap and water to reduce contaminating the dressing as it is applied.

- Stop severe bleeding with a clean cloth pad (preferably sterile) pressed against the wound. If the bleeding continues, add another pad and hold it tightly on top of the first. Have the student lie down and if the wound is on an arm or leg elevate the limb. Immediately send for help.

- After treating the injury, carefully and thoroughly wash your hands to remove any trace of blood. If you have any cuts or abrasions into which the blood might have passed, notify the health care worker who treats the injured student.

One of the most serious concerns when there is an injury where bleeding takes place is the potential for the transfer of blood from the injured person to the teacher or art activity leader. While it is never expected that students may be infected with blood-borne pathogens, it is possible, and when assisting an injured student, the teacher or art activity leader must exercise caution and, if possible, wear gloves. Information on what types of infectious diseases may be transferred by blood is extensive and "Exposure to Blood," a publication of the Centers for Disease Control, should be a required reference to include with other first aid materials.* Teachers or group leaders should be aware that if there is any indication of an infection in the injured student, that information must be kept completely confidential.

Burns

- First degree burns redden the surface of the skin. Cool the site with cold water and cover with a sterile dressing. Physicians often discourage the immediate application of burn ointments because the person may be sensitive to them and they may be difficult to remove later.

- Second degree burns cause blisters. Cool with cold water and carefully cover with a sterile dressing without breaking the blister.

- Third degree burns destroy the underlying growth cells and will appear white or black and leathery on the surface. Treat as other burns, if possible. For extensive burns, don't attempt to remove clothing. Treat for shock.

- Get the student to help as quickly as possible for second and third degree burns, especially if they cover a significant area.

- Flush chemical burns with large quantities of water; continue flushing until help arrives.

Shock

Shock is caused by a failure of the circulatory system and occurs after burns, emotional stress, or significant loss of blood.

- Make the injured person lie down, with head low and feet elevated, to maintain blood supply to the brain; keep him or her comfortably warm, but cover only if chilly.

- Call for medical assistance immediately.

* CDC on Exposure to Blood (www.davisart.com/safety)

Poisoning by Mouth

- Give water or milk in large quantities; include Ipecac syrup in first aid kits to induce vomiting.

- Look for antidote instructions on the poison container: follow them carefully and call a physician or poison control center immediately.

- Do not induce vomiting if the poisoning is from kerosene or other petroleum products or from any caustics such as acid.

This information is intended only to give some general insight into the first aid knowledge that is needed in art-making spaces. It should not be considered adequate instruction in proper first aid treatment. Nor are these injuries the only kinds possible in art-making processes. There is a great need for first aid information, and teachers are well advised to get assistance from medical personnel before it is required.

The Role of Regulations

As discussed previously in this book, there are few if any governmental regulations that apply to schools and virtually none that apply specifically to conditions in an artroom or other art-making spaces. This doesn't mean that information available from various agencies cannot be adapted to the art space and used to identify and correct problems. Rather, the lack of regulations makes it even more important for the teacher or group leader to seek out whatever general information is available about a problem or a potential source of hazard for students and determine how it can be used. Do not assume that because the art-making space does not fall under regulations it is safe to ignore potential problems.

NIOSH has prepared a series of Self-Inspection Checklists for schools and some of those have direct application to art spaces.* These, together with the suggested forms found in this book, will be useful tools in identifying and correcting hazards and potential problems.

---

* NIOSH Alphabetical Listing of Checklists (www.davisart.com/safety)

Chapter 3

Controlling Art Work Space Hazards

As is emphasized throughout this book, the art work space is a setting with unique problems resulting from the tools and materials used and the variety of processes performed. What is so unique about the art work space and what health or safety hazards are created? How can a teacher or art activity leader control the hazards inherent in some of these materials and tools so they can be used safely? It may never be possible to eliminate completely all the hazards, but they can be controlled and reduced to a point where injury or illness will be rare occurrences.

Proper Placement of Fixed Equipment

Most teachers and many art activity leaders have learned how important the distribution of supplies and equipment is to the success of an art lesson. Equipment also piques the curiosity of many students. If students are to be discouraged from exploring it and risking unnecessary injury, fixed equipment in particular must be placed so that the artroom doesn't become an "attractive nuisance."

Each art work space has different kinds of fixed equipment. Some, such as the electric kiln, are literally wired into one location and can't be moved. Others, like the paper cutter, drying racks, and freestanding tools, can be moved but seldom are. Teachers must decide where tools will be safest, and when the arrangement has been tested for workability and effective traffic flow through the space, the equipment should be left permanently in those places. Although there is no one proper configuration for placing equipment that is best for all situations, here are some useful guidelines.

Electric kiln location is usually determined by its wiring, venting, and by normal room traffic patterns. Fumes from firing must not, of course, be allowed to spread throughout the room and adjoining halls, so a canopy ventilating hood or down draft type of venting device is a requirement. Kiln venting is the major factor in locating the kiln.* Be sure the kiln is placed on a non-combustible surface. The kiln must also be placed away from usual traffic lanes and be separated from the students. Ideally, the kiln should not be located in the art-making space itself, but perhaps in an alcove or storage area. If it must be in the art-making space, a chain link barrier will keep students away from the kiln when it is hot and at the same time allow the teacher to monitor it visually.

Paper cutters are usually portable and moving them is often a way to clear a table top for other uses. This is a poor practice. If the cutter does not have its own stand, select a table (perhaps one discarded from some other use), shorten the legs so it is the proper height for the students to use, and attach the cutter to it. Put the cutter in an easily observable area of the room so that those who are permitted to use it can be seen.

There are two different types of paper cutters available: the familiar guillotine cutter with the arm that lifts and cuts on the down stroke and the rotary trimmer, which is the one most appropriate for use in art-making spaces.† The trimmer has a totally covered rotary blade, cuts more accurately and easily, and virtually all students can be successful using it. There are no statistics available as to how many students and teachers have been injured by a guillotine cutter (and most particularly, one on which the spring supporting the blade has been broken), but to use this type of cutter when a better and safer tool is available is not good thinking on any teacher's part.

Saws, drill presses, and other freestanding equipment are not uncommon in art-making spaces. These should be placed in little-used areas of the room to ensure a clear working space for cutting and drilling. Mark the floor around the machine with wide colored tape or painted strips to show the operator and helper where to stand and to indicate a limit behind which all others must stay. A table saw (rare in an artroom) must have a "safety zone" large enough to protect students from the occasional kick-back of wood improperly fed into the saw. It must be placed in a way

---

* AMACO Kiln Venting information (www.davisart.com/safety)
† Rotary Trimmer sources (www.davisart.com/safety)

that the direction of the blade rotation is away from the class. With proper instruction and supervision, these tools should pose no great safety problems for older students.

Printing presses probably will have to be moved out of the way when not in use, unless the room is used exclusively for printmaking. Attach a tabletop press permanently to a table of appropriate size, strength, and height. Store it either out of the artroom altogether or in an area where it will not interfere with student movement. Cover a press not in use to protect it from dust, to prevent it from being used as a storage surface, and, most important, to discourage students from playing with it. If possible, remove the spokes or drive wheel so that the rollers cannot be turned and to prevent students from inadvertently bumping into these spokes.

Drying racks should be located where there is good access and the shelves can easily be reached. There is little hazard with such racks except when they are improperly used for general storage—a function for which they were not designed and which may cause them to collapse or to become bent and not work properly. These racks can be used to help define room areas and should be located close to where they are used. Drying racks for silkscreen activities using oil-based inks should be vented to carry off the vapors from the evaporating solvents in the ink.

Properly located equipment can make the art work space a safer and more functional place to work. A clear understanding of how the equipment works and the amount of space each piece requires is needed to determine its location. How often and by whom it is used is the second factor in deciding location. This should not be done casually. Once you have determined a good location, equipment should remain there, but be prepared to move it if it does not work well.

Solvents

If solvents are strong enough to dilute or dissolve oil-based inks and paint or lacquers, it doesn't take much imagination to realize what they can do to human beings. Turpentine and lacquer thinner have been in general artroom use for so long they are often accepted as a normal part of everyday life. Yet these and other solvents should not be assumed to be safe, and the hazards they impose must not be overlooked.

As with other substances, the most effective precaution with solvents is to be sure they are necessary. Eliminate the hazard by eliminating its cause. With the availability of high-quality and versatile water-based ink and paint, the necessity for solvents other than water has been significantly reduced. Experimentation with acrylic paint should convince students that it is as much an "artist's medium" as is oil paint, and it can be used for almost any visual effect the student wants. The pigments in some paints may be hazardous, but if students keep the paint (and their brushes) away from their mouths and frequently wash their hands, the risks are minimal. Water-based silkscreen printing materials are very effective and eliminate the need for the extremely toxic solvents normally used in this process. Visual results with these materials may differ somewhat from those with petroleum-based inks, and on very large print paper shrinkage may cause registration problems. Nevertheless, water-based inks provide a genuine experience in all of the screen printing processes and are not hazardous to use.

Permanent type materials, which require solvents, should only be used in properly ventilated rooms, and they should not be used at all at the elementary or pre-school levels. Do not use paint stripper or varnish remover in the art-making space under any circumstances unless the label identifies it as nontoxic. These are probably the most dangerous of solvents and are completely unnecessary in any art-making program. Further, be extremely careful to use lacquer thinner only when absolutely necessary, for it too has very toxic contents.

Figure 3 provides information describing the solvents most often found in the artroom. These are reasonably safe if precautions are taken in their use, but there should be very little reason to need them in a well-planned program. Not to have them available would be the best overall decision.

Storing Dangerous Liquids

Solvents other than water are nearly always highly flammable. If they must be used, their handling and storage must be designed to reduce the possibility of fire. The following information is helpful in making decisions about using and storing these liquids.

- Liquids are considered flammable if they have a flash point below 100° F.

- They are considered combustible if their flash point is 100°–140° F.

Flash point temperatures listed for substances may vary slightly depending on the evaluating agency making these classifications (McCann, 1992, pp. 175–176), but since the variance is only about 20° it is simpler to use 100° as the figure differentiating combustible and flammable materials.

SOLVENTS OFTEN FOUND IN THE ARTROOM

Solvent	Hazard	Precautions
Paint Thinner (Petroleum distillates)	Inhalation may cause dizziness; irritating to eyes, nose, throat; causes dry, cracked skin. Affects mostly skin, eyes, respiratory and central nervous systems. Combustible.	Use sparingly; wash after use; keep container closed when not in use. Exercise care against indirect ingestion. Store in a proper cabinet. Do not use at pre-school or elementary levels.
Turpentine	Irritates eyes, nose, throat, and skin. Contact can cause sensitization (subsequent allergenic reaction). Kidney damage. Flammable.	Avoid use. Substitute turpenoid or paint thinner.
Lacquer Thinner (Toluene, ketones)	Causes dizziness, weakness, muscle fatigue, dermatitis. Inhaling large quantities can cause death. Attacks central nervous system, liver, kidneys, and skin. Flammable.	Limit use to specifically ventilated areas. Keep containers covered when not in use. Store properly. Do not use at pre-school or elementary levels.
Shellac Thinner (Denatured alcohol)	Poisonous. Irritating to eyes, nose, throat and is mildly narcotic in small amounts. Flammable.	Should not contain methyl alcohol. Use with close supervision but avoid use if possible. Keep container closed when not in use. Store properly. Do not use at pre-school or elementary levels.

3

Figure 3

Note: Flash point is defined as the temperature at which the liquid gives off enough vapor to form a mixture with air near the liquid's surface that will ignite when a flame or spark is present (McElroy, 1969, p. 1010). The lower the flash point, the more flammable the liquid. Flash point is not the temperature at which the material would spontaneously burst into flame, as is sometimes thought.

Flammable liquids (Class I) that might be found in the artroom are acetone, benzene, ethyl alcohol, toluol (contained in some lacquer thinners), and turpentine. Gasoline is also a Class I liquid. None of these are essential for art-making activities. **They should not be used or stored in an art-making space for any reason, because of their high volatility (the tendency to vaporize or evaporate rapidly), and for the health hazards they create.**

Combustible liquids (Class II), such as kerosene, mineral spirits, or lithotine, may be found in some art-making spaces. These are not as hazardous as Class I materials, but still should be stored in special containers and only used when absolutely necessary.

Storage Containers

Storage containers used for flammable and combustible liquids should be designed specifically for such use. The best container will be of heavy metal with a spring-closing lid. Figure 4 outlines maximum quantities of these liquids that can be stored in several types of containers.

Storage Cabinets

Storage cabinets for flammable and combustible liquids should be double-walled, built of 18 gauge metal, and have tightly closing, lockable doors. One-inch thick plywood with glued and screwed joints can be substituted, but doors should fit tightly and hinges should be sturdy enough to keep the doors from sagging. These cabinets are not intended to contain flames, but to keep fire from reaching their contents and should meet the NFPA (National Fire Protection Association) fire code. If that is not possible, have any cabinet used for this purpose approved by the local fire marshal. Clearly label the cabinet "Flammable—Keep Fire Away" and affix an appropriate official symbol to the outside of the cabinet to ensure that firefighters will know what is inside without opening the door. Open containers should never be left in any storage cabinet.

STORAGE CONTAINERS FOR FLAMMABLE AND COMBUSTIBLE LIQUIDS

	Flammable*	Combustible
Glass or plastic†	up to 1 gallon	1 gallon
Metal	1 to 2 gallons	1 gallon
Safety cans	2 gallons	2 gallons

3

* Flammable liquids should not normally be kept in the art-making space since there would be no reasonable use for them. (Among them are ethyl ether, acetone, benzol, benzine, ethyl acetate, ethyl alcohol, gasoline, methanol, and methyl ethyl ketone.) Only flammable aerosol sprays (a Class IA substance) might on some occasions be used in an art-making space.

† Always use non-breakable containers if possible. All containers must be labeled to identify their contents clearly. Keep containers covered except when actually in use to reduce the vapors in the air around the work area. Do not keep old, unused containers of flammable or combustible liquids in the art-making space; they should be properly disposed of and never left abandoned.

Figure 4

Waste Disposal

Rags or paper towels that have been used with flammable or combustible liquids should be disposed of in an approved waste container. These cans are constructed of sheet metal and have covers designed to open only partially. When released, covers will close automatically. These containers must be emptied at the end of every working day. The disposal of any waste materials or any waste solvents must be handled differently from normal trash, and these materials should never simply be dropped or poured into a regular wastebasket.

Disposal methods for solvent-contaminated waste will vary from school to school or from facility to facility where art is being made. It is the teacher's or art activity leader's responsibility to alert both the facility administrator and custodian to the fact that there is such waste and that it should be handled according to school or local government policies. Make this notification to the administrator in writing, indicating how much waste is likely to exist and how often it will need to be emptied.

Self-closing containers, approved cabinets, and acceptable waste containers are necessary if any combustible liquids are used. This equipment will significantly reduce the possibility of fire. However, real fire safety depends on consistent monitoring to be sure that proper procedures are followed. Again, always be sure nothing is kept in the art work space that is not necessary to the classes being taught there.

Adhesives

Scissors, glue, and colored crayons: there probably isn't an art-making space where they won't be found, even in places where there is no art program as such. They certainly rank as the most basic materials in art making. Of the three, glue is worth particular attention because it is so often taken for granted. Glue—or the more generic term, adhesive—comes in different forms and with different purposes, so there may be special problems in its selection and in supervising how it can be used safely in art making.

A wide variety of adhesives is available for use in art making, but the primary determinant of which you use is whether the adhesive is appropriate for the materials you want to fasten together. There is no point in using a glue that is stronger than what you are gluing together or in attempting to attach materials with glue that is not strong enough to hold them together. Read the labels carefully and find the glue most suitable for the work your classes will be doing.* There are many, many choices available and it is recommended that you explore the information on the container label before making a decision. Do not use rubber cement with preschool or elementary-age children or with classes of elderly or disabled students, and if it must be used with other groups, use it with great attention to the warnings that are found on the containers. Most rubber cements are highly flammable and toxic and should be stored in chemically suitable containers and used with excellent ventilation. Particularly, do not use any adhesive product that does not provide information as to its contents or warnings about its possible flammability or toxicity.

Scrap Materials

Art teachers and leaders of children's groups of all kinds have been brought up on a regular program of improvisation in their use of materials. Art from throw-aways has historically been the way to extend meager budgets and challenge the imagination. The practice is considered educationally sound, based on the creativity it takes to change trash into objects of beauty or visual interest. Sometimes it seems almost preferable to use junk in place of new materials. Contemporary interest in recycling even gives the process an environmentally positive twist as well. With such attitudes, there are probably few art teachers or art activity leaders who give more than a passing thought to the boxes and piles of "stuff" they have accumulated. But scrap should be as much a part of the regular inventory as are tempera paint jars and drawing paper.

Much of what teachers accumulate in the way of "supplementary" supplies is useful in making art objects. Wallpaper samples, corrugated cardboard, Styrofoam meat trays, scraps of cloth, yarn, wood, buttons, light bulbs, applicator sticks, shoe polish, feathers, bottle caps, mailing tubes, game parts—the list is long. The only limits seem to be the imagination and ingenuity, as well as energy, of the teacher or leader and the availability of storage space. In most cases these materials do not represent any unique health hazard or involve safety risks that are not obvious. The greatest problem is that their composition is unknown:

---

* Adhesives information (www.davisart.com/safety)

- If wood scraps are painted, what is the content of the paint? Is there a chance it is lead based, having come from old buildings?

- Was toxic dye used to color the feathers?

- Did the applicator sticks come fresh from a package or were they used for some unknown purpose before they got to the artroom?

- Who can guess what is in shoe polish other than various kinds of waxes, and what is known about them?

In general, there is no way to find out about any scrap materials. There is little possibility of tracing the materials, and it is unlikely there will be information on any label. Not knowing, then, means taking special care in their use:

- Wash hands frequently.

- Keep storage containers clean and free of dust.

- Keep all materials out of and away from the mouth.

- Do not heat or burn any of the materials (heat may release unknown fumes from paints or plastics).

- Handle sharp edges or rusty surfaces with care.

- Have students wear dust masks if there is any indication the materials are not clean.

Caution with Substitutions or Altering Materials

Both the use of scrap materials and the substitution of alternate materials are of concern. What are the implications of making those changes?

- When starch is mixed with tempera paint to make finger paint, does it affect the nontoxicity of the paint? How does adding liquid soap or glycerin to tempera paint (to adhere it to slick or waxy surfaces) change the paint?

- Some teachers may add a preservative to tempera to keep it from going sour or to paste to keep it from becoming moldy. Of these preservatives, phenol, for example, is highly toxic, while oil of cloves is only slightly so.

- Even if the preservative used is not a serious hazard, what happens when it combines with other substances? Is the character of that substance changed? Might there be a synergistic effect (two combined materials creating an effect neither has alone)?

Since the effects are not known, it is not advisable to make these or other alterations, even for the sake of economy. It would, at least, be wise to ask the manufacturer if they can provide information about what happens when these changes are made. The inventiveness of art teachers knows few limits, and sometimes that is a hazard in itself. Keep in mind that any certification of the nontoxicity of materials cannot be expected to hold true if they are altered by mixing them with other substances. Of what value is a toxic hazards identification program that may be undermined in this way? Working with scrap materials and turning discards into art is exciting and provides worthwhile experiences for the children, but teachers and leaders must be sure that unnecessary hazards are not brought into the art-making space with the scraps. As with all such problems, the teacher has two choices: not use the material at all or treat it with care and use unknown substances with respect.

Making a Materials Inventory

Controlling potentially hazardous materials is, of course, a primary goal in any art-making space. One of the most effective controls is to keep a complete inventory of all art materials used that are hazardous or potentially hazardous. An inventory provides a record of consumption, which helps determine how much should be stored in the artroom and to what extent those materials represent a hazard.

Any label information gathered in the inventory process can be used to set up materials categories. Label information divides materials into three categories: nontoxic materials, materials that are harmful under some circumstances, and materials whose contents are unknown. Use an inventory sheet similar to that shown in Figure 5 to record the following information:

- Type of materials in the room.
- Number or amount on hand.
- How often each is used.
- Where and how it is stored.
- Who has specific responsibility for it.
- What kind of disposal procedures are required.

Materials Inventory

TOXIC CONTENT CODE

Nontoxic   A
Toxic   B
Unknown   C

Room Number: _____
Teacher: _____
Date: _____

Quantity	Item	Description	Toxic Code	Notes/Date amount of re-order

Figure 5

This inventory provides a concise record of any materials that may cause health problems including allergies, as well as immediate information about the amount of all supplies on hand. If current, this inventory will be an excellent way to know the overall status of supplies and expedite ordering any materials. Its primary function, however, is to control hazardous substances.

Making this inventory should be an annual activity and will keep the teacher or art activity leader aware of materials that should not be used or need to be used with special care. Looking for information about toxicity should be a standard part of this inventory process; often the composition of art materials products change, and doing this inventory will refresh a teacher's or leader's memory about checking for changes in the materials they plan to order.

3

Chapter 4

Initiating Corrective Actions Through Knowledge and Cooperation

Naturally, the teacher or activity leader is the single most important factor in providing health and safety instruction and supervision. Without them committed to a health and safety program, even the most complete guidelines will be of little value. Such a program must be specifically designed for the conditions existing in each art-making space. Consistent instruction and enforcement of rules will inspire students with the same respect for art hazards. Prevention is the key: students should avoid materials that may make them sick and techniques that can injure them. With that as a fundamental premise, the following generalizations can form the basis for an effective health and safety program.

First, good art-making space management reduces accidents and injuries as well as health problems. Second, less toxic materials can usually be suitably substituted for toxic ones with little or no extra cost. Third, learning how to use tools and equipment correctly and safely is no more difficult than learning to use them incorrectly. And fourth, remember that practices learned in art-making activities will be those the students follow elsewhere and in subsequent years. Students are not just learning about art or how to make it; they are also learning how to use their tools.

Setting up a program begins with a thorough understanding of existing conditions; the information must be very specific. To get started, the teacher should list those existing conditions, carefully examine current materials and practices, and know student ages and health characteristics. Gathering this information may seem difficult and time-consuming, but it is necessary for the program to work. The outline of sample questions below highlights the type of information that must be gathered. Later chapters provide specific suggestions for gathering information and using it to make the artroom a safe and healthy environment.

General Art-Making Space Conditions

- Housekeeping: Is there dirt, debris, and dust? Is there adequate storage and access? Are there information and warning signs?
- Tools and equipment: What is their placement and condition? Is there a maintenance schedule for them?
- Lighting: Is natural and artificial light adequate?
- Ventilation: What are the provisions for general and local fresh air sources?

Current Practices

- Instructional methods: How and to what extent are health and safety included?
- Students' responsibility: What is expected of them? What is their responsibility based on?
- Monitoring procedures: Who ensures that correct practices are followed? What are the consequences of infractions?
- Art activity space management: How are the distribution and pickup of tools, the handling of hazardous materials, and the disposal of hazardous waste handled?

Age and Risk Group

- Bodily development: What are the natural defenses for the age of the students? Is manual dexterity well developed or has it been impaired by age or illness?
- Human weaknesses: Which students may have allergies, chemical sensitivities, or respiratory problems?
- Exposure accumulation: What is the frequency and term of exposure? Medical records: Is a medical record kept for the teachers and art activity leaders whose exposure is long term?

- Skill development: Do students have adequate knowledge of correct procedures or do they take things for granted?

Materials and Activities

- Toxicity: What are the relative ratings of materials used?
- Ingestion, inhalation, and absorption through the skin: What are the possibilities of these occurring?
- Power tools: Is noise level a problem? Are tools located or used with concern for safety?
- Activities in the curriculum: Have they been selected to minimize problems?

The teacher or art activity leader, being fully familiar with the art-making space and curriculum, is the one person who can compile this information, but help can come from the school nurse, the principal, the program administrator, teacher aides, other helpers, concerned parents, and students. Each of these persons has a stake in promoting a safe art environment.

This information itself does not create a health and safety program, but it provides the basis for one. Thus, when a teacher or activity leader is fully aware of existing conditions, the next step will be to implement appropriate practices. In so doing, a health and safety program is established.

The Students' Need to Know

The purpose of hazards instruction in the artroom is not to frighten any-one into feeling that only blunt pencils and plain paper pads are safe to work with. It is to call attention to the fact that art materials and processes, like many other things encountered in daily living, can be harmful if we are careless. An attitude of respect must prevail not only for the art objects themselves but for the way those objects are made and for the materials from which they are made. Being afraid to work with a material or tool makes little sense, and it is a teacher's or art activity leader's responsibility to provide instruction that will alleviate, not incite, that fear. The teacher is a role model for the students and, as such, must be especially careful to show respect for art processes and to demonstrate a sure knowledge of how to eliminate potential problems.

Students come to the art work space with no real understanding of the dangers involved with the materials or tools they will use. Even though some will be familiar with the art processes, most students are totally unaware these processes may involve problems that could affect their health or safety. Likewise, their familiarity with the process does not mean they know the proper methods for working with tools or handling materials.

Having studied the art-making space to determine where potential problems exist, find out what the students understand about those problems. To discover exactly what the students know about a process before they begin it, use a simple pre-activity test. Giving a pre-test for every activity covering what they know about the process, the tools, the techniques, and the hazards is not a terrific burden and can have an important place in art instruction. From the results, the teacher can tailor, for a specific group, instruction and expectations.

Teachers already use lectures, slides, instructional hand-outs, and demonstrations to introduce new activities. Finding out what students already know is another important part of that introduction. Some students may know the process from prior experiences and find another introduction boring; others may be so uninformed that typical methods of introduction will be totally inadequate.

A short pre-test will provide the following information:

1. Have they done this process before in another school or class?
2. Have they already mastered the skills required?
3. Do they know how to use the necessary tools?
4. Do they understand the possible hazards of the process?
5. Do they have any examples of previous work?

What are some specific art hazards questions that might be included in this pre-activity test? Figure 6 (page 43) is a sample test for a linoleum block printing project and can be used as a model. Completing this test several days ahead of a new activity will enable the teacher to devise the best introduction. It will provide valuable information for health and safety information. Large blocks of class time need not be spent on art hazards instruction if students already have a good understanding of the correct ways of using the materials and tools required for a planned activity.

Paperwork is already a big problem, particularly in schools. Teachers

What do you know about linoleum block printing?

Student name: _____

Date: _____

Teacher's name: _____

1. What is the purpose of using a bench hook?

2. How can you tell when a cutting tool is dull and should be replaced?

3. What is the reason for avoiding getting oil-based ink on your hands?

4. How should you handle the solvent you use in cleaning the block?

5. Where should you put dirty rags after you have used them for cleaning up?

Answers:
1. A bench hook keeps the block from slipping while it is being cut; it is the correct way to hold a block while cutting; it helps determine the direction cuts should go.
2. A tool is dull when it slips or slides across the block surface; when dull tools slip they can cause injury.
3. To keep the skin from having direct contact with solvents necessary to remove the ink; solvents can cause skin irritation or other problems.
4. Be sure to wear protective gloves; use the smallest amount of solvent possible to do the job; be sure the cleaning is done in a well-ventilated place or in the place the teacher indicates.
5. Solvent-soaked rags are very flammable and special self-closing cans should be used for disposal; fumes from dirty rags can be dangerous to breathe. Rags with solvent in them that need to be saved should be hung under an operating canopy ventilating hood to ensure the fumes are drawn out of the room.

Figure 6

4

are expected to keep many records. But in this case, the extra paperwork helps identify student understanding of a process and is worth the effort.

When students do not do well with the answers to these questions, special emphasis on hazards should be included in the general introduction-demonstration for the project. However, if they do know the answers, only general reminders along with any special information unique to conditions in the present art-making space will be necessary. Should it ever be needed, this type of written test will also serve as evidence that the students demonstrated they knew how to deal with the hazards of the activity.

Daily lesson plans should always include health and safety notes. These notes need not be extensive but will remind teachers of things the students should know about the materials they are using (e.g., "remind the class not to spray fixative inside the art-making space unless it is under an operating ventilation hood"). Awareness of hazards concerns will become second nature to both students and teacher if discussed on a regular basis. Lesson plan formats are usually either prescribed by the individual school or are a personal design of the teacher. Whichever is the case, planning for health and safety should be "built into" the form used.

Consider using the format in Figure 7 (page 45) as one way of doing this.

Student Committees

Most teachers or activity leaders know that the more students are involved in planning and carrying out learning activities, the more effective learning will be. This holds true for all subjects with any age level—health and safety instruction is no exception. Enlisting direct student participation produces positive results.

A health and safety committee for the school or program is one way to do this, but forming such a committee in each class is better yet. What does such a committee do? Its members can work with the teacher in identifying artroom hazards and help design ways to instruct other students in dealing with those problems. They may also make inventories of the materials in the art-making space, check label information about toxicity, discover which materials need special storage considerations, and establish procedures to guarantee that safe practices are followed. What these committees accomplish obviously depends on the age and maturity of the students, but beginning in pre-school and the primary grades, children can either volunteer or be selected to meet and talk about health and safety

1
2
3
4
5
6
7
8
9
10
11
12
13
14
15
16
17
18
19
20
21
22
23
24
25
26
27
28
29
30
31
32
33
34
35
36
37
38
39
40
41

Sample Lesson Plan Format

Grade/Class: _____ Teacher: _____

Activity: _____

Planned Length: _____

Lesson Objectives:			
Materials and equipment required for lesson:	Already have	Need to order	Safety and health hazards issues
•_____	n	n	
•_____	n	n	
•_____	n	n	
•_____	n	n	
•_____	n	n	
Motivation:			
Procedures (continue on back):			
Evaluation:			

Figure 7

hazards. Perhaps the discussion with young children would involve only how scissors should be handled and used safety, but with older students more complex issues can be discussed:

- What does the word *toxic* mean?
- How can we find out about the toxicity of materials?
- What do we do with materials when we don't know if they are toxic or not?
- Should we make rules about working with art materials?
- How will we know the rules are being followed?
- What will the penalties be for not following the rules?
- How can we make health and safety rules known to the rest of the class?

Students and activity leaders can decide what materials or practices should be limited or not allowed. Students can help devise ways to monitor and enforce rules and can also work with the program or school administration in solving critical problems. Encourage your students to shoulder some of the work involved in health and safety education. Have them involve other students in an important aspect of curriculum planning, which will increase learning all around. They will also be useful in suggesting to the school or program administration changes or improvements in how materials are purchased or in the physical conditions in the art-making spaces. Involving the students in this way also demonstrates that the teacher is fulfilling a significant responsibility in the health and safety aspects of the art-making experience. Obviously, health and safety education cannot be left entirely to student committees, but the program effectiveness will be enhanced by actively involving students in the process.

Part Two

Understanding the Nature of Health and Safety in the Art-Making Space

Chapter 5

What Constitutes Health Hazards?

Defining Toxicity

The word *toxic* is particularly difficult to pin down. Loosely defined, toxic means poison, but that is overly simplistic. One definition of toxicity, provided by the EPA, is "Deleterious or adverse biological effects elicited by a chemical, physical, or biological agent."* However, consider this definition, which has, perhaps, broader implications: "[Toxicity is] a relative property of a chemical agent and refers to a harmful effect on some biologic mechanism and the condition under which this effect occurs" (McElroy, 1969, p. 1325). The key parts of this definition are the word "relative" and the phrase "condition under which this effect occurs." This means, essentially, that toxicity is not absolute and what might be fatal for one person in one set of circumstances might not be for another. Stated in yet another way, "In practical situations the critical factor is not the intrinsic toxicity of a substance per se but the risk or hazard associated with its use. Risk is the probability that a substance will produce harm under specified conditions" (Doull, 1980, p. 12).

Children under twelve and older adults are "high-risk" groups because their physical development may increase their vulnerability. But, exposures causing serious reactions in some persons do not affect others as much or at all. These are examples of the relativity of toxicity and explain why just knowing the toxicity of a particular substance is not totally sufficient to safeguard students.

* EPA Terms of the Environment: Glossary (www.davisart.com/safety)

Toxicity rating information is extremely important, but knowing students, knowing the duration of exposure, and knowing the extent of the exposure are also important to a complete understanding of the possible harmful effects of materials. Above all, know whether there are any conditions or materials found in the art work space that will cause harm if the teacher, art activity leader, or students are exposed to them.

Knowing Your Materials

Too often, it does not occur to teachers that art materials may cause health problems. When a student—not always in an elementary grade or of preschool age—spreads the glue on the paper with his finger and then licks off the residue, the teacher's reaction is usually to say, "Don't do that!" But that reaction is probably more because the act is socially offensive than from concern that it might make anyone sick. Sometimes, however, putting some kinds of materials into the mouth, inhaling their fumes, or even touching them can have serious results. For example, teachers should know that:

- Wallpaper paste often contains toxic preservatives; read the label to be sure it is safe to use.

- Vapors from some kinds of permanent felt markers are often toxic and repeated exposure can result in serious gastric and nervous system damage as well as liver damage; however, many permanent markers are now safe to use if they carry a label stating they are nontoxic. Be sure to check.

- Batik wax vapors are flammable and can ignite at temperatures within the range of an ordinary hotplate; overheating also releases irritating fumes.

- Nearly all the chemicals used in photographic processing can cause skin, eye, lung, or respiratory tract irritation or allergic reactions.

In selecting materials for the art-making space, the teacher or art activity leader has several obligations concerning the potential health problems. First, learn as much as possible about the materials used in art processes and know whether they have any inherent harmful qualities. Second, using a form such as that in Figure 8 (page 51), find out whether students have any special health problems that might be aggravated by exposure to these materials or allergic reactions to the substances they

Notice of Need for Information Concerning Allergies and/or Respiratory Problems

During this year, we will be working with a variety of materials and processes in our art classes. We make every effort to ensure that the materials are not harmful to students, but occasionally someone has an allergy or special sensitivity to something we do not know about. Will you please provide the information requested below so that we can avoid any unnecessary problems?

Student name: _____

Date: _____

My (son) (daughter) (self) has known allergies: Yes n   No n

My (son) (daughter) (self) is allergic to the following substances:

_____

_____

My (son) (daughter) (self) has known respiratory problems: Yes n   No n

If I or my son or daughter am/is exposed to a material that has never before caused an unsuspected allergic reaction, please call me or my physician for instructions:

My phone number: _____

My doctor's phone number: _____

Signed: _____
  (Responsible party)

Figure 8

might be using. For children, have a parent sign the form and provide the information; for adults, have the responsible person (usually themselves) sign and provide the information. And, third, carefully and correctly instruct students in the use and handling of the materials. Obviously, the teacher's overall supervision in the art-making space must ensure that all materials are used as they should be.

Label Information

There is little reason for teachers and art activity leaders to be concerned with the health impact of materials that are not specifically used in the art-making space. Broad knowledge about the hazards of art materials certainly can be useful in making decisions about new activities, but in most cases it is not critical. It may actually be distracting in the first stages of identifying art hazards. Go through the supply cabinets and be sure samples of everything there are set out for examination. Crayons, tempera paint, watercolors, felt markers, pencils, adhesives, solvents, inks, and all other products need to be looked at. Include an example of each material from each manufacturer, and spend enough time making notes from the label information to get a good understanding of what is there.

Art supply cupboards often yield some very old, long out-of-use products. They should be discarded both because they may have deteriorated and because label information may be inaccurate. Sometimes old materials can be dangerous. For example, a product called Fibro-Clay, produced by Milton-Bradley many years ago, contained asbestos. When research disclosed the cancer risks of asbestos, the product was immediately recalled, but it may still be (unbelievably after so many years) in some artrooms and may yet be used by a teacher who discovers it in the back of the storage area and is unaware of its problems. Many art teachers and activity leaders have been conditioned not to dispose of any art materials that might someday be useful. Although it may seem unlikely that some are saved as long as they are, be sure to check everything.

Usually one of two different types of label information is available, and each presents some problems in interpretation.

Labels with Little or No Information

This type of label creates a most difficult problem in evaluation, because some teachers assume that if no warning is stated, there is nothing to worry about in using the material. Not so. Manufacturers have not always been required to disclose contents, and art materials are often stored for years before they are used.

Repackaged Materials

Another concern is that materials have sometimes been repackaged and no information from the original label survives. Materials without label information are potentially even more hazardous than those with the most specific warnings. This is true because there is no sure way to know of the possible dangers that may be associated with their use.

Mislabeled Materials

Few obstacles to obtaining accurate label information present more difficulty than that of mislabeling. This can occur when supplies have been repackaged or when a label has come off one package and has been re-applied to a different package. This would be most likely to occur with clays or glazes, but it could happen with any packaged material. Mislabeling will be obvious with some materials (e.g., a mislabeled container of paint), but being attentive to whether the contents of a package match the label information is important. If you have reason to be suspicious that the label on the package does not apply to the contents and there is no way to determine what the correct labeling should be, the material should be discarded.

Dealing with Inaccurate or Unreliable Content Information

When little or no content information or warning is provided, or if it is obvious repackaging has occurred, it is best to discard the product or, at least, to treat the materials as hazardous. In general, be guided by the following:

- Do not use any unlabeled materials with elementary-age or younger children.
- Use unlabeled materials—particularly clay and glazes, the materials most often repackaged—with older students only if there is good general room ventilation and if hands are protected or washed frequently.
- Keep all such materials covered when not actually in use.
- Store the unlabeled or repackaged materials carefully and in small

5

quantities in places where they won't get knocked to the floor, broken, spilled, or inadvertently opened and used.

- Instruct students not to put the material into their mouths.

Whenever there is a question about the safety of using materials for which the label is inadequate and does not provide toxicity information, the teacher or art activity leader should look for a "Material Safety Data Sheet" (MSDS), which is a standard form used in industry to make statements about their product's safety.* With the increase of clear and unequivocal nontoxic statements on most of the art materials now available, MSDSs are not needed as much as they once were. For those who feel the information they have is incomplete, these forms are, for the most part, readily available online, along with information on how to interpret them.

Labels with Nontoxic Statements

Within the past decade major legislation and standards development have resulted in significant changes in the usefulness of health labels on art materials. The major work in ensuring proper labeling has been done by the Art & Creative Materials Institute, Inc. (ACMI), of Hanson, MA. Their booklet *What You Need to Know About the Safety of Art & Craft Materials,*[†] which explains the process by which the toxicity of art materials is determined, is extremely useful reading for the teacher or art activity leader. Figure 9 illustrates the ACMI certified labels.

In brief, ACMI has established testing protocols to determine if a product meets the chronic hazard labeling standard ASTM 4236 (American Society for Testing Materials)[‡] and the U.S. Labeling of Hazardous Art Materials (HAMS). Additionally, ACMI testing is used to determine the quality of some art materials and some products may carry its label, which indicates that the product has met the quality standards that have been developed by the American National Standards Institute (ANSI).

Quality standards, which do not affect the nontoxicity of art materials, have not been fully developed, so some products you may be familiar with will carry labels from agencies other than ACMI, such as ASTM or ANSI. The ACMI publication discussed in this section as well as their *Certified*

---

* MSDS source reference (www.davisart.com/safety)
† ACMI booklet source (www.davisart.com/safety)
‡ ASTM reference (www.davisart.com/safety)

THE ART & CREATIVE MATERIALS INSTITUTE LABELS

Products bearing the new AP (Approved Product) Seal of the Art & Creative Materials Institute, Inc. (ACMI), are certified in a program of toxicological evaluation by a medical expert to contain no materials in sufficient quantities to be toxic or injurious to humans or to cause acute or chronic health problems. These products are certified by ACMI to be labeled in accordance with the chronic hazard labeling standard, ASTM D 4236 and federal law P.L. 100-695 and there is no physical hazard as defined with 29 CFR Part 1910.1200 (c). Additionally, products bearing the CP (Nontoxic), CP Seal, or AP with Performance Certification meet specific requirements of material, workmanship, working qualities, and color described in appropriate product standards issued by ACMI or other recognized standards organizations. This will be indicated by the accompanying wording "Conforms to ASTM D 4236 and ANSI Performance Standard Z356."

ACMI *Certified Products List,* 2003, inside cover.

Products bearing the new CL (Cautionary Labeling) Seal identify products that are certified to be properly labeled in a program of toxicological evaluation by a medical expert for any known health risks and with information on the safe and proper use of these materials. This seal appears on only 15% of the adult art materials in ACMI's certification program and on none of the children's materials. These products are also certified by ACMI to be labeled in accordance with the chronic hazard labeling standard, ASTM D 4236, and the U.S. Labeling of Hazardous Art Materials Act (LHAMA).

ACMI *Certified Products List,* 2003, inside cover.

Figure 9

Products List can be searched to determine if a product you intend to buy has met their certification standards for nontoxicity.*

The ACMI program has since 1940 evaluated art, craft, and other materials. Their scope has been expanded and the program now has certified over 60,000 craft and creative materials formulations of art and craft materials for both children and adults. This certification program is perhaps the most important there is for art teachers, art activity leaders, artists, craftspeople, and anyone who finds her- or himself making art. By knowing about the program and understanding the meanings of the labels, all persons making art can be protected from hazardous materials.

How Materials Affect the Body

Acute injury provokes obvious body reactions. Cuts usually bleed; a smashed finger causes immediate pain and swelling and, before long, a blackened fingernail; touching a hot stove reddens or blisters the skin. In an even more dramatic example, a person swallowing poison will be immediately sick and may even die. Action is taken immediately because symptoms are seen or felt.

But not all hazards are this obvious. What happens internally may be hidden and often does not produce any noticeable symptoms until it is too late to take corrective action. There are many familiar examples of this: lung cancer and emphysema, which often develop after long periods of smoking; cirrhosis of the liver, which can be brought on by alcoholism; and, even more common, periodontal disease, which develops because of inadequate mouth care and bacteria removal. By the time the symptoms of these serious illnesses are recognized, the condition is sometimes beyond help. These are known as *chronic* reactions to toxins.

Art teachers should understand how chemicals from art materials can enter the body, what happens when they do, and what can be done to eliminate or reduce acute or chronic effects. With repeated exposure over time, some substances accumulate in the body. Teachers and art activity leaders who subject themselves to these materials on a regular basis should be particularly concerned. While students may work with these materials for a few hours a week for several months, teachers usually work with them many hours daily for many years.

Even though there are no immediately visible effects, it can be assumed that if a hazardous substance is found in the artroom, anyone coming into

* ACMI Certified Products List booklet (www.davisart.com/safety)

contact with it will suffer some damage. The only sure prevention is to entirely eliminate contact with the substance. Knowing how these chemicals enter the body and the extent to which they are harmful will help in making decisions about whether to continue their use in the art program.

There are three ways foreign substances enter the body, and each provides access to vital organs and systems. Although the human body has a marvelous and complex defense mechanism, it can be breached and overloaded. It is the overload that must be prevented.

Skin Absorption (Contact)

The skin is an important part of the body's defense, but it can break down with exposure to certain substances, such as acids, caustic materials, solvents, and bleaches. This is especially true when cuts or abrasions are present. Rashes, blistering, itching, or flaking of the skin can result. Damaging substances can enter the bloodstream through the skin and quickly move on to other parts of the body. This is particularly true with older adult students whose skin has so changed with the years that it offers little or no protection.

Some art materials may be sensitizers which, in certain persons, initiate allergies that did not previously exist. The reaction to these materials can become increasingly severe with each subsequent exposure. Although desensitization shots sometimes help, there appears to be no sure way to reverse this effect. To ensure these problems do not occur, it would be necessary to avoid any contact with such substances. If, however, the materials must be used in some essential processes, hands should be protected either with gloves that are resistant to the particular chemicals or with barrier creams, which help shield the skin for up to four hours. These creams must always be re-applied after hands have been washed and care should be taken that the correct type of cream is used for the materials being handled.

Inhalation

Large particles of inhaled dust will be trapped, to a great extent, in the mucus of the nose or in the upper respiratory system. However, fumes, vapors, gases, and most fine dust particles are not usually stopped by this "first line of defense," and some of them (those from overheated plastics, from glacial acetic acid used as a stop bath in photography, and from welding, for example) can irritate and damage the lining of breathing passages and lungs. More complete information concerning specific respiratory reactions to various airborne chemicals can be found in Spandorfer, Curtis, and Snyder (1996, pp. 6–7) and in McCann (1992, pp. 46–49). In general, however, prolonged exposure may cause bronchitis and emphysema, which are aggravated by the effects of tobacco smoke, auto emission fumes, and other pollutants in the air.

Some dusts, such as free silica in clay powder, can lead to silicosis or pulmonary fibrosis. Powder forms of fiber reactive (cold water) dyes or vapors from turpentine and epoxy hardeners may cause respiratory allergies or asthma.

Generally, it is possible to avoid using most of these materials without restricting the art program, but some, such as clay, must be included. Short of total elimination, anything that creates airborne fumes and particles must be controlled. It is imperative to plan for a separation of the mixing and working activities, to use correctly selected dust masks, and to provide good ventilation. (Adequate ventilation requires special considerations, which are discussed in Chapter 2.) At the very least, however, cans of solvents, inks, and adhesives must be kept closed when not actually in use, and floors need to be frequently wet-mopped; sweeping or dry mopping raises dust and does more harm than good.

Ingestion

Ingestion extends beyond the obvious example of young children eating glue. It more often occurs indirectly any time the mouth is touched by hands or tools carrying contaminants. A pensive moment, when fingers unconsciously touch the lips, is enough to transfer traces of some materials. "Pointing" a brush with the mouth is another example of indirect ingestion.

Except for swallowing vast amounts of material (an unlikely accident in an art-making space), most substances enter the body through absorption of very small amounts in the stomach. The bloodstream transports contaminants to the liver for detoxification or to the kidneys for filtering. But harmful substances can overwork and damage those organs (as alcohol causes cirrhosis of the liver). The body then cannot cleanse itself of its own toxic products, to say nothing of foreign substances. Remember, the effects of many substances are cumulative and therefore over time can be quite serious. Even small ingested amounts must be avoided. Simple but regular practices such as washing hands, keeping them away from the mouth, and not biting fingernails (to avoid picking up what may be lodged under them) will reduce possible ingestion. However, the only really effective control is to be constantly aware of how students work and to be alert for any incidents of mouth contact with art materials.

Health and Safety Instruction and Supervision

The best way for students to learn to use materials correctly is to give clear and precise instructions when they first encounter any materials in the art class. This is especially true if students have previously adopted careless work habits. Allow ample time for this instruction when planning lessons. It is inefficient and ineffective to hurry through health hazards instruction in order to get into the project work more quickly—students will be unlikely to retain much, and the necessity for frequent reminders later on will interrupt the flow of work time.

Knowing about materials, being aware of student health problems and applicable health and safety issues for the age of the students, and enforcing proper work habits are key to safe art-making space procedures. There is no substitute for these in the effort to protect students from harmful interactions with art materials and processes. Figure 10 provides a brief summary of necessary health and safety information based on the age of the student.

The Difference Between Safe-to-Use Materials and Safe Use of Materials

Remember, regardless of the age or skill level of the students, that having safe materials is not the same as using them safely. Nontoxic, for example, does not mean a material can be used in ways it has not been designed to be used. Mixing materials or using them carelessly can create safety problems quite apart from the problems or health issues that were considered when the products were labeled nontoxic.

SUMMARY OF HEALTH AND SAFETY INFORMATION BY AGE

Age	Health	Safety
Pre-k–12	Because of the extensive labeling of art materials, it is only important to check and be sure what is being used is nontoxic and that the product carries the appropriate labels to verify that. Any ceramic materials used must be checked carefully for toxicity and kilns must be appropriately installed and used.*	These early years are when safe practice habits are instilled in children and thus this is the most important element in a health and safety program. Thoroughly review the most effective and safe way to distribute materials and to use equipment and be sure instruction is accurate and consistent. (Review Chapters 2, 3, 5, 6, and 7.)
13–18 and most adults	Increasing the kinds of art experiences this age group works with requires much more careful attention to the toxicity of all materials used and the appropriate use of protective clothing and equipment when working with various processes and materials. Thoroughly review information about the nature and use of all materials and the most effective procedures for using them. Be aware of the need for proper ventilation and the safe installation of equipment.	Thorough instruction in the safe use of tools and equipment is essential as the complexity of processes increases. Regular review of this information with students is critical. Classroom organization and use of space are basic to safe practices and the condition of all tools must be regularly monitored. Awareness of potential problems is critical. (Review chapters in Part Two, especially information about ceramic kilns.)†
Seniors	This group of students, like the pre-k–12 students, is considered at high risk for injury or illness brought on by unsafe use of materials. Be sure that all materials carry a nontoxic label and do not substitute unknown materials for those that are properly labeled. Brush use in painting must be carefully monitored to be sure no transfer of even nontoxic materials takes place (especially to the mouth). Ceramic kilns must be properly installed and vented and should never be used unless they are.	Even more than a group of younger adults, this age group may be very set in the ways they work and it will be difficult to change their practices. Be sure instruction on procedures is clear, understood, and followed; do not overlook the possibility of hearing or sight impairment. Hand tremors can provoke accidents, so be aware of this issue and prepare for it. Limit the equipment used unless there is a clear need and demonstrated knowledge of its use. Review the chapters in Part Two and apply that information to the students in the class.

* L&L Kiln Mfg.: Kiln Installation (www.davisart.com/safety)
† EPA Ceramic Kilns (www.davisart.com/safety)

Figure 10

Chapter 6

Sources of Safety Hazards

The Safety Environment

The general conditions existing in an art-making space set the stage for careful or careless work habits. To overlook these conditions is to overlook the basis of many hazards. Adequate space, appropriate storage, and good housekeeping produce the safest working conditions. Sloppy and cluttered work areas do more to cause accidents than almost anything else, but the effective management of all materials, equipment, and work in progress will overcome this problem. Each of the following areas should be evaluated regularly.

Floors

Floors are often overlooked in art-making space safety. Normally art-making spaces are on one level. But if the room is shared with other disciplines or is an old room that has been assigned as art work space, this is not often the case. This is particularly true for art making in non-school settings. There may be changes in level such as steps, platforms, or risers or the floor itself may be uneven, especially in made-over spaces often used for adult classes. If so, the edges of steps should be clearly marked with contrasting color for good visibility. Because changes in level are not usually under teacher control, traffic paths should be established well away from them if possible. Be aware that irregular changes are more likely to cause tripping or falling than just the change itself. In any case, it is important to reduce the use of any area where such changes occur.

Floors should be kept dry and non-slippery and anything spilled on them should be cleaned up at once. For instance, spilled water around the sink should not be allowed to puddle and should periodically be wiped dry during each class period. Students of any age should accept this responsibility, but they should also know when to call for the teacher's or activity leader's help; pre-school-age children in particular should be watched carefully.

Custodians or building owners should be notified when floors develop cracks, loose or broken tiles, or torn or frayed carpeting, any of which can trip people. If woodworking or wood sculpture projects are done in the room, be aware that wood chips and sawdust can increase the chance of slipping, especially on wooden floors and very smooth concrete. Keep floors clean by using a damp mop or shop type vacuum cleaner. Avoid sweeping, especially when students are in the room, since it raises dust and usually succeeds only in rearranging much of the waste materials; sweeping compounds are generally expensive and ineffective in controlling dust. If the floor is particularly slippery, especially around equipment or in worn places, report the condition to the school principal, building owner, or program director. Request that it be treated with a paint preparation containing abrasive particles or covered with non-slip adhesive strips where students stand or walk.

Plumbing

For some processes, particularly any that require acids or solvents, emergency washing facilities should be close to the working area. If a student is splashed with these substances, it is necessary to flush the exposed parts of the body with water immediately. To do this effectively, the proper equipment must be available; in some cases, this will mean an overhead shower. Often, however, a sink equipped with a flexible hose and spray attachment is an acceptable substitute for a drench hose since the chance of a serious splash in most art work space is slight. Eyewash fountains or water faucets with inexpensive eyewash fittings should be located in any areas where acids are used. If there is no water source in the immediate area, portable eyewash stations containing at least five gallons of water will serve the purpose. They must be checked regularly for contamination of the water and changed as necessary for it to remain fresh. Such equipment should be at the proper height for the students who might need it.

An often overlooked plumbing hazard is the hot water temperature at the sink. If the temperature is erratic, students can unexpectedly be burned by scalding hot water. If this occurs, it should be reported to the custodian or building owner, and warnings and reminder signs at the sink may be necessary if the problem cannot be corrected.

Lighting

Certainly, all areas of the room must be well lighted, but no specific amount of light can ensure safety. Generally, work surfaces with a dull, matte finish will improve visibility and reduce glare. Some equipment, such as a jig saw, requires local, direct lighting, which should be placed so as not to be distracting or shine in the operator's eyes.

Adequate light not only improves visibility in working areas but helps to prevent tripping over unseen objects. Poor lighting cannot be justified on the basis of energy conservation; such false economy is not worth the risk of injury. In work spaces that are poorly lighted, night classes may need to be eliminated. Supplementary lighting may be useful, but be sure lamps and extension cords are not located where students can trip on them. Be aware also of the limits of the electrical circuits and do not overload them.

Room Size

Activities should properly correspond to the size of the art-making space. It is poor policy to expect students to work at any art activity unless the space is completely adequate. Overcrowding should not be permitted and for classes held in non-school settings, this is often a major problem. School and program administrators sometimes consider it "efficient" space utilization to confine a relatively small art class to a small room, but the trade-off may well be the unnecessary injury of a student. Curriculum planning which calls for activities that crowd the space available is also poor policy. If the room size cannot be adjusted, the curriculum has to be.

Furniture should be arranged so students can work without being impeded. Many tools and some materials require space for arm swinging (as with hammers) and broad, sweeping movement (as with wire manipulation). The arrangement of participants needs to be organized accordingly. Be sure too that space between work tables is kept clear of obstructions and allows for easy movement through the room. The movement of students and materials or equipment around the room should be limited and carefully planned. Be sure also that the room arrangement allows for an easy exit in the event of an emergency.

6

Storage

Adequate storage in the artroom should be provided for bulk supplies, materials in use, and partially completed projects. Providing this storage space prevents cluttering and allows for easy movement. Contending with a crowded working area will promote careless work habits among students and constant reshuffling of materials to clear work space is annoying and can cause impatience, which may lead to accidents as well as damaged materials.

Art teachers and many art activity leaders are notorious collectors of materials of all kinds that may prove useful for future projects. No one advocates arbitrary or wasteful disposal of such materials, but these "collectibles" must be neatly stacked or boxed and stored in areas completely out of normal traffic patterns or work areas. A periodic culling of such materials should take place—any materials that have not been used in a year probably never will be. These materials, boxed or not, need to be periodically dusted so that when they are used, the dust they have accumulated will not be an irritant to any student.

Noise

Power tools that are used frequently will create a disturbing level of noise in the room. It may be necessary to restrict the times they may be used and to require students working with them to wear hearing protection. Earplugs or industrial type earmuffs will significantly reduce the potential for hearing loss.

Keeping Order in the Art-Making Space

The myth that mess, sometimes called "creative chaos," makes for better art has permeated many artrooms and art-making spaces. It results in the frequent loss of tools and equipment tossed out by accident, or waste of materials, such as torn or dirty paper or contaminated and dried-up paint. It also makes for very unsafe working conditions.

Actually, there is little reason to believe creativity is enhanced in any way by a messy work area. The irritation of having to search for tools lost amidst clutter is hardly conducive to exciting and innovative art work. Nor does the waste and carelessness seem likely to have any positive effect on the quality of art. In fact, the free and ready access to tools and materials characteristic of an orderly environment probably enhances the free atmosphere of the work area and allows for greater creativity.

Sometimes these "ordinary" safety hazards are overshadowed by the more "exotic" problems such as getting rid of chemical fumes. But it isn't very sensible to worry about adequate ventilation and at the same time forget that serious injuries can occur from tripping over materials that are obstructing traffic. The fact that the carelessly stacked cans of paint that have crashed down on someone's head contain nontoxic paint is of no consequence to the injury that might result in such a happening.

There are several obvious reasons for maintaining the art work space in good order. Some points have to do with safety, others with economics, but all relate to developing and sustaining an atmosphere that encourages art production. Be aware that:

- Properly stored tools remain in good condition longer, and tools in good condition are less likely to cause accidents.

- Dirty floors tend to be slicker than clean floors. Bits of paper, and particularly sawdust, reduce traction and increase the chances of slipping, especially on waxed floors.

- Carelessly stacked materials invite problems; it sometimes becomes necessary to pull paper from the bottom of the stack and the results of that are obvious to anyone who has tried it.

- Trying to work on litter-strewn tables or desks reduces effective control over tools and equipment, leading to careless actions. School books, coats, and especially backpacks should always be put well out of the way of working or walking areas.

Keeping an orderly artroom isn't very exciting, but it is absolutely necessary in any art-making space safety program. Doing so demonstrates to students that proper care of equipment and material is the key to a good working environment. If no litter and dust are allowed to accumulate, it is easier to find effective methods of controlling other hazards. For example, keeping the clay storage and mixing room in meticulous order reduces dust control problems—wet-mopping can then take care of most of the dust, and less of it has to be exhausted by mechanical devices.

The teacher or art activity leader must have a positive attitude toward an orderly working environment and storage areas; fortunately, it is one that doesn't have to wait for special budget action. Perhaps even more important, it shows a concern that will help convince administrators when there is need for special equipment to combat other types of hazards. When a teacher shows that everything possible is being done to eliminate hazards, administrators of any program are more likely to provide funds to do the rest.

Working with Tools

Clear instructions must be given when the equipment is first used, and those instructions must be followed. Correcting the student when the equipment is being used improperly and helping the student develop the right habits are also part of the teacher's and art activity leader's role. The simplest tool has the potential to cause serious injury, and it is often familiar tools that cause problems. What are the functions of the most typical tools and what determines necessary instruction?

Cutting

Most tools in the art work space fall into this category. Scissors, paper cutter, saws, knives, wood and linoleum cutters, metal snips, and wire cutters are all common, and all require special initial instruction along with frequent reminders about correct use.

- Do not assume the students know how to handle and correctly use any cutting tools, even those with which they ought to be the most familiar.

- Give careful instruction and supervision for blunt scissors just as for pointed ones to ensure good habits for all types of scissors. Do not assume that blunt scissors are any less dangerous than pointed ones because their previous use may falsely give young children the sense that all scissors can be handled in the same way the blunt ones can be.

- Use the proper tool for the job. Do not, for example, cut wire with metal snips or scissors, cord or string with wire cutters, wood with linoleum cutters, or cardboard with the paper cutter. To do so dulls or damages the tool and makes it unreliable when used for its proper purpose.

- Keep cutting tools sharp. Dull or chipped blades slip easily and cause painful cuts. If necessary, have sharpening done professionally, or discard the tool—it is false economy to continue using a tool in poor condition. It is good economy to teach students how to sharpen tools themselves.

- Store tools so as to protect them and in a way that makes distribution easy and safe: scissors in scissor holders, knives in drawers, saws and metal and wire cutters in racks.

- Set up controlled distribution procedures that minimize handling of tools.

- Establish specific work spaces for any power tools.

- Severely limit and personally supervise student use of the paper cutter and be sure the blade spring is always functioning. If possible, substitute a rotary trimmer for a guillotine cutter; injuries are far less likely to occur.

- Be sure that any material being cut is held securely: vises and clamps should be available when needed and students should know how to use them.

Piercing

Although not done as often as cutting, piercing materials is even more hazardous because it involves sharply pointed tools such as punches, awls, drills, scissors, and compasses. Puncture wounds to hands, fingers, legs, or eyes are not uncommon in piercing activities. Careful supervision is at all times necessary. Hand drills are probably the least dangerous piercing tools to use, provided they are sharp so as to reduce slipping and used with the correct bit. A vise to hold the material is absolutely necessary if the drilling is done into a rounded or irregular surface.

Pounding

Hammers and mallets should always be used appropriately. Obviously, wood or rubber mallets shouldn't be used to pound nails, and steel hammers shouldn't be used to hammer print linoleum blocks. But this has more to do with damage to the tools or to the work than with safety. However, damaged tools will not function well when they are used later. Be aware, too, that hammers come in different weights. They should not be too heavy for the student or the job for which they are used.

Squeezing and Pinching

The tools used in these processes are not especially dangerous, providing reasonable care is exercised to be sure fingers are not caught in them. Pliers, vises, and printing presses are usually the only tools used for these purposes, and simple instructions should be sufficient. However, a printing press usually has rollers and a moving bed, and instructions should be particularly clear about these danger areas on the press.

Heating

Processes such as batik, wax encaustic, printing on fabric, welding, or soldering require hotplates, irons, or butane or other types of torches. These are potentially very dangerous both because of the possibility of direct burns and because they are sometimes used to apply heat to very flammable materials. Use low temperature and double boilers for melting waxes, and closely supervise and control the use of torches. Be aware of the flash point temperature of wax used for batik; use candy thermometers to maintain a safe margin, and do not exceed that temperature. If hot wax does catch fire, do not try to extinguish it with water, because that will cause major flaming of the wax. Cover the container and remove it from the heat immediately. Burns are among the most painful and disfiguring injuries. Every effort must be made to prevent them and to know how to treat them if they do occur. Welding tanks must be chained to prevent them from falling over and rupturing or damaging the gauges; welding hoses must be kept clear of any regularly used areas where they might cause tripping.

Power Tools

Power tools are increasingly common in art-making spaces: the prices of electric drills, drill presses, band saws, saber saws, and sanders are more and more within the reach of many art budgets. The same precautions concerning the use of handsaws and drills are appropriate for power equipment of any type:

- Keep all tools in good condition.
- Use sharp blades and bits and store the tools properly with the bits and blades removed.
- Carefully instruct in and supervise the use of the tools.
- Designate specific places for power tools to be used.
- Use three-pronged, grounded extension cords, if they are required. Use cordless, battery-powered tools whenever possible. Completely disconnect freestanding tools from the power source when not in use.

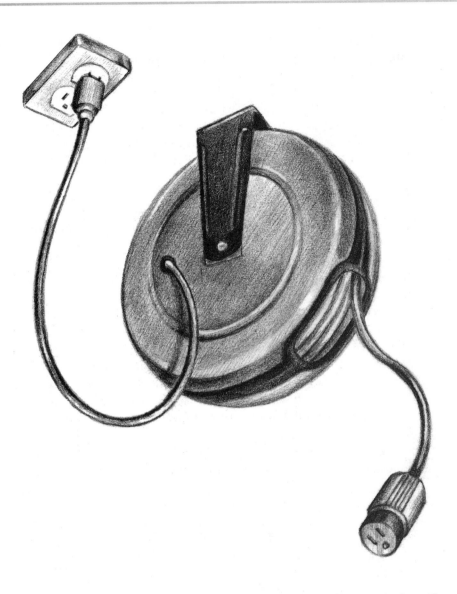

Extension cords and hoses should be placed so they are not tripped over, causing hot spills or pulling things from tables. Electric cords on retracting reels, hanging above the work area for easy access, are a good solution to the problem of cords on the floor.

6

There may be other tools and equipment, which have not been specifically mentioned here, and each teacher or leader must see that students know, understand, respect, and properly use any of them.

Extension cords and hoses should be placed so they cannot be tripped over, which could lead to injury from hot spills, pulling things from tables, or a fall. Electric cords on retracting reels, hanging above the work area for easy access, are a good solution to the problem of cords on the floor. However, use cordless, battery-powered tools whenever possible.

Art-Making Space Safety Check

Checklists are the most effective way to control hazards. While most teachers may rightly feel that their teaching time is already seriously eroded by bookkeeping tasks of dubious importance, the time taken to make a conditions check can be very worthwhile. It is also a good way of involving students in a learning experience.

Safety checks should be regularly scheduled each week and, in schools, rotated through the different classes to give all students the experience. If the students participate as safety "reporting monitors," it soon will be apparent to them that room and tool conditions are important. They will learn what to look for in identifying hazards. When they have learned to do this as a regularly scheduled activity, it will become second nature for them to notice and report any conditions that may seem hazardous. Awareness of safety conditions is best learned by the regular experience of monitoring those conditions. When a teacher demonstrates that safety is important enough to give it time and attention, the students will recognize it too. This is a valuable life-long attitude that will develop through this simple, helpful, and regular art-making space activity.

Use a form similar to the Safety Conditions Check (Figure 11) for recording the hazard information. It should include sufficient space to ensure all important conditions can be noted and provide a way to indicate clearly when something needs to be corrected.

When a hazard is found, the teacher should correct it if possible or call it to the attention of the principal, the program administrator, or the custodian. By recording and reporting hazard information on a specific form, action is much more likely to occur than if the notification is made by an informal note or a casual comment in the hallway.

Not all aspects of safety can be easily checked by students. Certainly conditions of the room as well as tools and equipment within the room can be done by students, but what about safety as it relates to procedures,

Safety Conditions Check

Room number: _____

Teacher: _____ Date: _____

	O.K.	Needs attention (comment)
General		
Floor condition		
Lights		
Aisles clear		
Storage area		
Ventilation		
Windows open easily		
Exhaust fans working		
Tools		
Hand tools in good order		
Power tools in good order		
Electric cords		
Tool storage		
Materials		
Well-organized storage		
Solvents properly stored		
Added comments		

Submitted to administrator:

Teacher signature: _____ Date: _____

Acknowledgment: _____ Date: _____

Figure 11

work habits, or the thoroughness of safety instruction the teacher gives? Since there are no "safety examiners" who can evaluate teacher performance, it will be very difficult to get reliable observations. The task thus falls to the teachers themselves. Several points should be reviewed on a regular basis:

- Are safety notations included in all lesson plans?
- Are regular and consistent instructions given for working procedures and tool use, even for tools frequently used?
- Are the students' work habits monitored? Are corrective comments made and is appropriate behavior demanded?

To determine how well the students are learning, evaluate and record the overall manner in which they work by using a form similar to the Performance and Attitudes Toward Safety (Figure 12). When students are working on an activity in which specific safety procedures are to be followed, use this form to record how they work. It is a good idea to check them from a location in the room away from usual observation points; stand at the back or side of the room so as to have a fresh perspective and reduce the chance of overlooking details. In addition, it may be helpful to ask the principal, the program administrator, or a fellow teacher or a teacher aide to use the Performance and Attitudes Toward Safety form and to share with you their observations of the class. This not only provides an outsider's view but also helps these people become aware of safety practices in your art-making space. Having such a visitor should not be seen as a threat and is a way of being sure that important safety practices are reflected in the way the students work when observed by others.

Responding to Uncooperative Behavior and Failure to Follow Instructions

Each teacher or art activity leader in his or her art-making space should develop procedures for dealing with uncooperative students and know what must be done so that the class is not disrupted. Of particular importance is a plan for dealing with these students when it comes to the health and safety rules and procedures the students are not observing. In some ways, failure to follow health and safety rules is more serious than any other type of uncooperative behavior. Because illness or injury can be the result of behavior or attitude problems it is essential that the consequences

Performance and Attitudes Toward Safety

Room number: _____

Teacher: _____ Date: _____

	O.K.	Needs attention (comment)
Performance		
Work area orderly		
Tools used correctly		
Materials well organized		
Distribution efficient		
Collection well done		
Tools properly stored		
Eye protection worn as needed		
Hair kept out of way		
Instructions understood		
Attitude		
Tools well cared for		
Attention to work (no horseplay)		
Concern about work habits		
Reminders necessary		
Comments		

Teacher observer's name: _____ Date: _____

Teacher initials: _____ Date: _____

Figure 12

of misbehavior are clearly established and clearly conveyed to students. No matter how talented the student might be and no matter how high the quality of his or her work may be, failure to follow established procedures for the safe use of materials and equipment must be cause for immediate cessation of work.

There can be no exceptions and no second chances in a specific work period. It is only in this way students can understand and internalize the importance of the health and safety rules. If, as is discussed in Chapter 4, students are involved in developing these rules, problems should be minimized because they know and understand their importance. But lapses must have their consequences and the Health and Safety Plan for the class must be followed in your art-making space.

Chapter 7

Hazards of Working with Common Art Materials and Processes

Not all art activities are hazardous enough to warrant special attention. However, many important art activities do involve materials and processes that present unusual problems in health and safety. Some use chemicals or tools that are especially hazardous. These are "special hazards" and need to be well understood.

All of these activities are important to any comprehensive art program and should not be discontinued because of their potential danger. Rather, the problems should be recognized and minimized. Students should be properly instructed so they can benefit from the art experiences yet not be subject to risk. With concern and careful planning this is certainly possible.

Drawing

To most teachers and group leaders, drawing seems to be one area where there are few problems of health or safety. Usually this is true. For most students, working with crayons, markers, pencils, pen and ink, charcoal, brush washes, or pastels are activities that cause few problems. (Obviously, attention must always be given to the handling of sharp tools such as pencils and pens to be sure they are used safely.) Some exceptions exist:

- The most serious problem is from aerosol spray fixatives used for charcoal and pastel, which should be used only with extremely good ventilation or preferably outdoors. Using spray fixative in the hall outside the artroom is not a solution, since that may contaminate the entire building. Also, charcoal and pastels usually create large quantities of dust in the artroom, which can irritate some students.

- Only water-based markers or markers clearly identified as nontoxic by the ACMI label should be used.

- Check oil crayons carefully. Many are imported and the nontoxic label may not be entirely accurate. Use only oil crayons that carry the ACMI nontoxic label.

Painting

Regardless of the type of painting done in the artroom, some pigments should always be suspect. While most pigments are nontoxic, there are a number of toxic inorganic pigments which should be avoided. A brief listing of some of these pigments is found in Figure 13. Other good resources to consult include *Making Art Safely* (Spandorfer, Curtis, and Snyder, 1996, p. 83) and the more extensive information on pigment toxicity included in the Health and Safety in the Arts searchable online database.*

Many of these toxic pigments are corrosive to the skin and cause irritation of the respiratory tract and mucous membranes. They also produce allergic reactions. Precautions in the use of these pigments will reduce problems and so will good housekeeping, keeping food out of the work area, and maintaining careful personal hygiene (frequent hand washing, for example). Brushes should never be put in the mouth.

Unless a class or group of students is working with these specific pigments, the greatest hazards in painting will come from turpentine or other toxic solvents if they are used. Obviously if the painting medium is acrylic, watercolor, or tempera where the solvent is water, that hazard doesn't exist. The use of oil paint can be included in a school or other art program (especially with adults and older adults), if certain precautions are taken. Ventilation is critical to ensure that the air in the work area will not be a hazard. Even more important, in recent years solvents, such as odorless Turpenoid, have been developed that are nontoxic and can make oil painting an acceptable activity for high school and adult students. Oil painting should not be done at the pre-school, elementary, or middle school level.

Teachers and group leaders should also remember that there is little that can be done in oil paint that can't be done equally as interestingly in acrylics. While it is now possible to use nontoxic solvents with oil paint,

---

* Health and Safety in the Arts Searchable Data Base: Painting Pigments (www.davisart.com/safety)

Pigment	Toxicity
Naples Yellow (Antimony)	Moderate to high; Reproductive toxin
Cobalt Violet (Arsenic)	High; Suspected carcinogen
All cadmium pigments (Cadmium)	Moderate to high; Reproductive toxin
Chromium Oxide Green, Viridian, Chrome Yellow, Zinc Yellow, Strontium Yellow (Chromium)	Moderate to high
Cobalt Blue, Cobalt Green, Cobalt Yellow, Cerulean Blue, Cobalt Violet (Cobalt)	Slight to moderate
Flake White	High; Reproductive toxin
Naples Yellow, Chrome Yellow (Lead), Manganese Blue, Raw Umber, Burnt Umber, Mars Brown, Manganese Violet (Manganese)	Moderate to high
Vermillion, Cadmium Vermillion Red (Mercury)	High

Figure 13

acrylic paint, with its faster drying time and easy cleanup, should not be avoided and, generally speaking, there is very little to be learned about painting through the use of oils that acrylics cannot provide.

Printmaking

Both health and safety hazards can be significant in printmaking. Careful instructions and procedures will minimize them. Health problems include exposure to inks, solvents, and acids, while the safety concerns relate primarily to injuries resulting from the use of cutting tools and the crushing action of various parts of the printing presses.

The health problems are most easily solved at the elementary printmaking levels because hazardous materials and processes can be eliminated; the safety hazards are most easily solved at the advanced levels by simple but thorough instructions. Safety must be the major target at the elementary level, since children are still learning skills required with tools. At the secondary level, health hazards resulting from exposures to the chemicals in inks and solvents are difficult to deal with and may present serious problems for students who encounter them.

Beginning Printmaking

Found objects, vegetables, glue, cardboard, and linoleum are the materials most used to make prints; tempera, finger paint, and water-based block printing ink are most often used to print them in the most introductory print courses. Safety hazards here primarily involve the tools used to cut the surface of the image carrier or block. As always, there is no substitute for good planning, careful instruction, and continuous monitoring of the way students are working. To minimize problems:

- Limit beginning students to working with relief or simple stencil prints. There is a sufficient variety of these experiences to provide many print-making opportunities without attempting other processes. Water-based silkscreen can be used with middle school students, but these students should be sufficiently advanced in the silkscreen process to justify the cost of these expensive materials.

- Be sure all tools fit the age, size, and muscle control of students. Cutting a design into the slippery face of half a potato with a paring knife is no task for most children, and probably not for many adults either.

- Show students how to hold their work as well as the cutting tool so that when the tool slips (as it will now and then), no harm is done.

- Sharp tools must be used so that cutting is sure, constant, and predictable.

- Cutting activities must be kept separated from the printing area to reduce confusion, mess, and contaminating the printing inks.

- Linoleum blocks must be cut only with linoleum cutting tools, and these tools must be used only for that purpose; never use razor blades.

- Meticulous instructions must be given for any type of press, and children should work in pairs during the printing steps so that one can monitor as well as help the other.

- Improper practices must be immediately corrected.

- No horseplay can be allowed in the vicinity of cutting activities.

To keep health hazards to a minimum, be sure only water-based, non-toxic paint or ink is used. There is no need for solvents other than water with these inks; in fact, there is never any reason to use oil-based inks with children under twelve or even with older adults. The problems associated with oil-based materials and their solvents are far greater than any advantages of permanence or color quality. Block printing on fabric, which requires oil block printing inks or special textile inks or paint, should be delayed until middle school or senior high and done only with proper precautions and gloves and under adult surpervision.

Print processes offer a great variety of learning experiences and expressive possibilities, so printmaking should be encouraged and exploited. With thorough planning, good instructions, and constant monitoring, the potential hazards of all of these processes can be effectively controlled. Because printmaking is a fairly complex activity, students of all ages can, at the same time, learn important lessons about responsibility so that hazards encountered later in the more advanced activities will be better understood and significantly reduced.

Advanced Printmaking

In middle school and senior high school or in classes with adults, many printmaking processes are the same as those in the elementary grades. Only the content and images become more complex, and skill levels increase. To this extent, the same hazards exist for older students and can be handled in the same manner. However, permanence of the print may become more a part of the design concept (textile printing or posters to be displayed outside, for example), or new processes such as silkscreen or intaglio printing may be introduced. To make permanent block prints, oil-based inks and the solvents they require are necessary. While the hazards of these may be reduced by good housekeeping practices and careful instruction, they cannot be completely eliminated. The problems, of course, are that these oil-based inks and solvents produce harmful vapors, which must be ventilated; can contaminate the skin, which must be protected; are flammable, so they must be properly stored; and produce waste materials that require special disposal.

To eliminate these problems, consider using water-soluble silkscreen inks, which are permanent and are also available in fluorescent and textile inks. These materials require only soap and water for cleanup. While silkscreen produces visual effects very different from block printing, in general it is possible to develop designs that will be just as effective. The reduction of the hazards involved is worth the extra effort that will be required. Water-based substitution for intaglio processes is not possible

7

and all precautions to protect the students from ink, solvent, and acid exposure will have to be taken.

Solvent and Ink Fumes

The only effective way to control vapors is through proper ventilation. Most schools do not have adequate local exhaust systems for extensive printmaking programs and cannot afford them. As an alternative, keep solvent cleanup activities centralized to an area where air can be moved away from the students by a fan and open window system. Remember, however, this requires incoming air to make up what the fan blows out (and relying only on open windows also would always make printmaking a warm weather activity). Keep all solvent containers closed except when actually being poured, and do the pouring in a well-ventilated area and use a nontoxic solvent such as Turpenoid whenever possible. Immediately put the solvent- and ink-soaked rags and paper into a closed waste container to contain the fumes. Strictly limit the number of students working in the print area at one time so that minimum amounts of ink and solvents are in use. If the odor of the solvent permeates the general art-making space, the ventilation system is not working effectively—either the volume of air being moved must be increased or the number of students working at one time must be decreased. See Figure 4 and Chapter 2 for possible solutions.

Contamination by Skin Contact

This problem can be significantly reduced by using latex gloves (be sure to ask students about allergies to latex prior to using latex gloves) during all phases of the printing process. Gloves may seem clumsy at first, particularly for students with small hands. But with time, using them becomes quite natural and virtually eliminates the problem of ink or solvents coming into direct contact with the skin. Latex gloves are not as impermeable as once thought, so they need to be changed frequently and immediately at the first sign of tearing or leaking. The type of glove will determine its effectiveness. Almost any household plastic glove will protect from ink contamination. Gloves used with solvents must be more carefully selected and should not be expected to provide long-term protection from heavy solvent contact. If the use is limited and the gloves are replaced frequently, significant protection can be expected. Careful disposal of dirty gloves, as with waste rags and paper, is a must.

Disposal of Waste Materials

All solvent-soaked rags and wastepaper must go into self-closing covered waste cans, and these cans must be emptied daily. Whether the contents of these cans represent a sufficient quantity to require special waste disposal procedures is a matter for the school principal or program director to determine in consultation with the building owner and the appropriate local health officials. The teacher should inform the administrator that waste materials are being generated as part of regular art activities and request instructions about proper disposal. Waste solvents should not simply be washed down the sink drain, since that usually leads directly to city water treatment plants and can create serious pollution problems.

Unquestionably, there are problems to be solved in order to work with printmaking safely. However, these can be largely dealt with through careful management, attention to procedures, thorough instruction, and a plan of action for all involved—teacher, activity leader, and students. The visual rewards of prints are such that teachers should have strong motivation to work out any problems and enforce their solutions. At the same time, this will help students learn the health and safety practices that need to be instituted. They are likely to carry these lessons with them throughout their lives. It isn't easy, but the results justify the effort.

Fibers and Dyes

In general, the hazards involved in fiber work are not extreme, but two aspects of fiber work require some attention if health hazards are to be eliminated.

Bacteria, Dust, Fibers

The first of these is perhaps easier to handle; it involves working with the carding, spinning, and weaving of various fibers. In most schools, yarns are purchased commercially, and concern about bacterial contamination is virtually unnecessary, since these materials can be expected to be free of problems. However, when wool or other fibers are used in carding or spinning, be sure the raw fiber has been thoroughly sanitized. Contamination by improper sanitization is rare but has occurred and can be very serious. It is a problem simply avoided. Do not work with any raw wool, for example, which has been brought into this country "informally," and use only products that you can be sure have been properly sanitized.

7

Working with jute rope or burlap in weaving or macramé projects can produce very irritating airborne fiber or dust particles that may aggravate or even cause respiratory problems in some students. Wearing an inexpensive dust mask will reduce these problems significantly.

Dyeing

The second aspect of fiber work that creates potential health hazards is working with dyes. There are a great variety of dyes available, which are used in different processes and for different fabrics. Unfortunately, not a great deal is known about the potential problems associated with them. Recent studies have shown that some food dyes, previously believed harmless, are now thought to be carcinogenic. Benzidine congener (family) dyes, used in many common products, have also been found to be carcinogenic and have been discontinued in some applications. Many other dyes and the chemicals used with them as mordants (particularly chrome, ammonia, and oxalic acid) can cause various toxic reactions, including respiratory and eye irritation, skin corrosion, and allergies. According to Spandorfer and colleagues (1996, p. 87), "Glauber's salt (ammonium sulfate), urea, and vinegar are the safest mordants." Part of the difficulty is that dyes are often repackaged by distributors, and warning information that appears on the original packages is sometimes not transferred to the new container. Fiber-reactive (cold water) dyes seem to be the most hazardous, causing symptoms such as asthma, hay fever, swollen eyes, and, after long exposure, sudden severe allergic reaction (McCann, 1992, p. 487).

In all work with dyes, it is recommended that a dust mask and gloves be worn to reduce exposure. When mixing dye powder, use an entire package at one time so that leftover packages will not spill and release dust. McCann (1992) also suggests that when possible, the entire dye powder package be submerged in water while it is being opened to prevent any inhalation. He also suggests making a glove box. Shellac the inside of a cardboard box to seal it, put a glass or plexiglass sheet on the top to see through, and cut hand holes in the sides. Wear gloves and mix the dye inside the box. No mask will be needed, and there is no messy cleanup required.

Tie-dyeing and batik are relatively common dyeing processes carried on at nearly all art activity levels. Common household dyes are often used in these processes, and they should be handled with great care. Wearing gloves is very important. Thoroughly wash any parts of the body on which the dyes have accidentally been spilled. The process of mixing powdered dyes is difficult to control and should be assigned only to students with the proper training.

Batik involves the use of heated wax, which is highly flammable and can cause painful burns. Use only a double boiler setup, where the wax is placed in a container that, in turn, stands in water to which the heat is applied. Never heat wax directly on a hot-plate, which might accidentally be turned to an unsafe high temperature. In any dyeing process requiring the heating of the dye bath, exercise extreme care to prevent scalding.

Textile Printing

Both block printing and silkscreen printing on textiles are popular activities with many students, from middle school students to older adults. These processes allow a variety of craft activities, perhaps the most common of which is printing on T-shirts. For permanence, either an oil-based block printing or silkscreen ink is often used, although these inks have the disadvantage of making the fabric stiff. Special textile inks, which leave the material soft, are often preferred. In either case, the inks themselves, and the solvents required to work with them, create the same toxicity problems as in other print processes. Whether nontoxic Turpenoid or a commercially prepared thinning oil is used, precautions must be taken to reduce skin contact and inhalation of the fumes. This means use of gloves or barrier creams and good ventilation of the work area. The use of acrylic paints or inks is recommended as a substitute that eliminates most problems. Manufacturers of acrylic products may have suggestions for using their products in textile printing. Don't assume ink or paint must be petroleum-based to be permanent.

As with so many processes, good housekeeping practices and common sense are fundamental to eliminating hazards associated with fiber work. In many cases, the problems mentioned here will not arise. Working with commercial yarns is the norm in most classes, and dyeing is usually a very limited activity. Exercising reasonable care, wearing dust masks and gloves when they are appropriate, and being meticulous in instructing students about working procedures are usually sufficient. If more advanced processes are desired, the hazards of both dust and dye should be thoroughly researched with the intent of devising procedures to protect everyone.

7

Stained Glass

There are several glass fabricating processes generally referred to as stained glass, but only one of these, leaded glass, seems likely to be done in an artroom setting. Working with epoxy resins or concrete to fill the channels between very thick, faceted glass, or laminating glass panels into layers are processes more likely to be undertaken by artists specializing in this medium and should be done in schools only with very advanced instruction. Leaded glass, however, is well within the skill sets of students at the middle school and senior high levels and certainly all adults. Attention needs to be given to those aspects of the process that can cause injury or long-term health problems.

Glass Cutting

It looks incredibly easy when done by an expert, but cutting glass often doesn't turn out quite as it should. At best, clean cuts leave very sharp edges that need to be smoothed with emery paper. Obviously, the glass must be handled with great care to avoid cuts. When the planned cut is less smooth than intended, and the trimming must be done with grozing pliers, extreme care should be taken in handling the glass. Not only are the edges of the glass likely to be jagged, but the small pieces and slivers of glass must be watched very carefully and contained as much as possible. Work should be done over a cleared, smooth, and clean surface so the fragments can be easily brushed into a waste container. Goggles must be worn to protect the eyes from any shards of glass that fly in unexpected directions. As each piece is cut, set it well out of the way so that it will be handled as little as possible before the assembling of the project.

Handling Lead Came

The major precaution necessary in handling the lead came, which holds the glass pieces together, is to be sure that hands and fingernails are thoroughly washed afterward. There is little problem with skin contact, but particles of lead, from cutting or sanding the came, can be passed to the mouth and ingested. If enough is ingested, the typical problems associated with lead poisoning may result. Keep hands clean and wipe the work surface frequently with a damp cloth.

Soldering

The solder used in leaded glass work is a 60-40 or 50-50 mixture of lead and tin and should not have an "acid core." In the soldering process, it is the solder that melts, not the lead came, and the lead fumes given off from that solder should not be inhaled. Good general ventilation in the work area plus a fan that will blow the fumes away from the participants should be sufficient, but a better method is to rig a vacuum cleaner with its intake near the work area with the exhaust hose arranged to carry the fumes outside. If possible, work in front of an open window with an exhaust fan to extract fumes (McCann, 1992, pp. 162–173). Do not solder under a canopy hood, since the fumes will be drawn up and pass directly by the student's face as he or she works. In general, wear gloves that will protect hands from being cut by sharp glass edges, yet still allow free manipulation of the glass. Limit cutting to simple shapes until good cutting skills have been achieved. Exercise care in the process of applying the putty, so that sharp edges of the came, particularly at joints, do not cause cuts.

Because of the manipulative skills required to cut or handle glass and the harmful cumulative effects of lead fumes, leaded glass work should not be undertaken until middle school or senior high or with adults.

Ceramics

In recent years, ceramics has become an increasingly popular art form. There are probably few schools and other art-making settings where ceramics activities don't occur and many where the program is extremely sophisticated and extensive. It is not only one of the most pervasive art activities in our society, it is also one of the most complex. It involves a number of often ignored hazards.

From the raw material to the finished object, clay goes through several different states and a series of manipulations, each of which has inherent health or safety hazards. It is also in three of these stages the most serious hazards occur: mixing, glazing, and firing.

Clay Mixing

Normally, the digging and refining processes occur before teachers and students come into contact with clay, so the first problem usually encountered is with clay dust that escapes from the bags in which it is packaged. During commercial preparation, clay is ground very finely so that leakage

7

from these bags is not uncommon. The bags sometimes are ripped in delivery, but more often they are torn open after arrival and partially used. As a result, the environment in which clay is stored is almost always a very dusty place. This clay dust is then scuffed into the air and tracked throughout the art work space, leaving a film of powder over nearly everything in direct contact with it.

Silica may compose up to 60 percent of the clay. Usually it is chemically bonded with other elements, but, if not, it is known as free silica and can be the cause of chronic silicosis (Seeger, 1982, pp. 14–15). Silicosis is ultimately a disabling disease of the lungs. Serious problems usually require exposure over a period of ten to fifteen years, and, therefore, the ceramics teacher and serious student are in the greatest jeopardy. For their protection, as well as for short-term students, correct clay handling and mixing procedures must be maintained. The teacher's own risk should be sufficient motivation to ensure an effective program of dust containment. No one should be unnecessarily exposed to free silica.

In addition to the health hazards, the safest possible procedures for mixing clay must be devised. Pug mills and clay mixers should be used only after detailed instructions have been given and tested and always with close teacher supervision. Incidentally, be sure that barrels used to collect clay for re-use are frequently checked for foreign materials or small tools that may have been inadvertently dropped in. These could later create serious problems in the mixing equipment.

The following suggestions should be observed to limit clay mixing hazards as much as possible:

- Determine if the quantity of clay to be used warrants mixing it from powder. With most students in most settings, the quantity needed does not justify exposing the students to clay dust hazards. While it is somewhat more expensive, ready-mixed clay is preferable because it eliminates the need for dust handling or mixing equipment. An explanation and demonstration of how clay is made will be sufficient until students are working at a more advanced level.

- Store and mix clay in an area separated from the studio so as to reduce the area of dust contamination.

- Keep all powdered clay bags in storage covered tightly with polyethylene sheeting in order to contain the dust.

- Stack clay bags off the floor on pallets or shelves so that cleaning the floor is easier and more complete.

- Wet-mop the floor of the mixing area frequently; never sweep or dry-mop, since this stirs up dust particles and provides inadequate cleaning. Vacuuming is effective only if there is an HEPA (high-energy particulate air) filter in the vacuum cleaner, which will prevent recycling of the dust through the air.

- Wear a dust mask specifically designed to filter out silica and other particles whenever working in the clay mixing room or when sanding a dry, but unfired (greenware), object. Be sure the mask is the correct type.*

- Have a local exhaust system operating whenever the clay mixer is in use; this will draw off most though not all of the problem particles.

Glazing

The second stage of clay work involves the process of glazing. This is a complex area, since the great variety of materials used in mixing glazes makes it difficult to generalize about the hazards. Glazes are often purchased ready-mixed, but in some instances the teacher or art activity leader will have the students learn glaze formulations, using bulk glaze chemicals to mix their own colors. In either form, simple rules should govern their handling:

- Allow no food in the area where glazes are being used; it is very easy to contaminate that food and thereby ingest chemicals.

- Always use a stirring stick; never use hands to mix the glaze.

- Wash hands thoroughly after working with the glaze.

- When applying the glaze, work on a surface that can be easily cleaned; Formica or plastic are suitable. Clean with a wet sponge or rag when finished.

- Glazes should be sprayed only in a booth with an exhaust fan that effectively carries away excess airborne particles.

When mixing the glazes from powder, exercise extreme care to contain the dust. This means there should be effective local ventilation or that toxic dust masks or respirators must be worn during the mixing process.

___

* Mask source information (www.davisart.com/safety)

Protective clothing such as a smock or shop coat should also be worn during this process. To contain dust, wear these garments only in the mixing area; wash them frequently.

Glaze Materials That Should Not Be Used

Some glaze materials are suspected carcinogens or are highly toxic and, regardless of how carefully they may be handled, should not be used. While it might be possible to control them in the studio of a professional potter or even the studio of a very limited number of advanced college students, there is no justification for subjecting even secondary level students or students in adult classes to the hazards inherent in them. Material that should not be used include:

- Lead and its compounds: -acetate, -silicate, -bisilicate, -monoxide, -oxide
- Arsenic and its compounds: -oxide, white oxide, -trioxide
- Cadmium and its compounds: -oxide, -sulfide, -chloride
- Nickel and its compounds: may produce highly toxic nickel carbonyl in firing
- Beryllium and its compounds: -oxide, beryl, beryllia
- Zinc chromate
- Selenium and its compounds: -oxide, -dioxide
- Any uranium compounds*

Perhaps the most recognized of glaze hazards is lead, although it is sometimes used in art-making spaces. The usual justification for this is that the objects being glazed will not be used as food containers, and thus there will be no chance of contamination. Two factors need to be considered, however. First, if any other objects are fired along with lead-glazed pieces, they are likely to be contaminated by lead fumes (as is the air around the kiln during firing). Thus, the lead can be passed along in sufficient quantities to represent a hazard to ultimate users of the objects.

Second, lead frits, often thought to be non-soluble and therefore safe for use, may not always be so. "Lead frits vary in solubility (the capacity to dissolve in liquids to form a solution) depending on factors like: method of

---

* Barazani (1978, pp. 19–20); Seeger (1982, pp. 27–35).

production, size of particles. Due to the variation in solubility some inhaled and ingested particles may dissolve in body fluids. Therefore, we believe that lead frits are not reliably nontoxic and should be handled with the same precautions used for other lead compounds. . . . Copper used in, or on, ware fired in a kiln with lead containing objects can affect the solubility of the lead on other ware" (Seeger, 1982, p. 32). Although there may be some justification for professional potters to use these materials in their private studios, there can be no such justification for their use in an art-making setting involving children or older adults.

Firing

The final step in the ceramics process, which involves hazards to teachers, art activity leaders, and students, is kiln firing. It involves intense heat around the kiln, and the vitrifying clay or glazes give off a variety of fumes, some of which may be highly toxic. All kiln firing, for example, produces carbon monoxide when various impurities in the clay decompose during firing. Depending on the clay and glaze content, gases such as sulfur dioxide, fluorine, chlorine, and nitrogen oxides are produced. Fumes are also produced by the heating of any metal above its melting point (Seeger, 1982, p. 41). Because of the number of chemicals involved, and the variety of possible mixtures, identifying a few general guidelines to reduce hazards is not really possible. Ceramics teachers, therefore, have a special obligation to keep up with the technical literature relating to safety. They should continually question manufacturers about health and safety hazards related to their products. Some important questions are:

- What is the exact nature of the chemical changes caused by firing glaze substances? What fumes are likely to be released and how can protection from the most hazardous be provided? Are there chemicals that must not be mixed because of a synergistic effect in the firing?

- What is the best location for the specific kiln(s) in use? If they are inside the building, what is the volume of air (cfm) necessary to ensure fumes are carried off through a ventilation device? How far should a kiln be located from a flammable wall surface?

- What is the maintenance schedule to be followed for each specific kiln type and kiln part? Who is qualified to carry out maintenance? Do the kilns require inspection by a fire marshal? What type of periodic inspections should be done and by whom?

7

- What safety controls are available for each kiln to prevent overfiring or missed shutoff times?*

Clay is a wonderfully creative medium in which to work. Following the general precautions presented here should prevent the conditions that can make clay an unsafe medium. The full-time or specialist ceramics teacher, who is probably also a serious potter, will need to constantly search for information from sources such as NIOSH (Centers for Disease Control and Prevention), OSHA (U.S. Department of Labor), and the EPA.†

Photography

Photography is an important part or even at the center of many art programs. Once the high school darkroom was under the control of the English department and used only for developing the pictures used in the school yearbook. Now the darkroom is more likely a facility central to the entire art program. With the production of new and relatively inexpensive equipment, including good quality 35 mm cameras as well as digital cameras, many students will have formal training in photography in the seventh grade or earlier.

Pinhole cameras and film developing are not uncommon experiences even in the third grade. Of all processes, film photography is the most dependent on chemicals. It is, in fact, totally a process of chemical manipulation, from the first exposure of the film to the printing of the picture. Consequently teachers and art activity leaders who direct students in photo work should know, for example, which chemicals may cause some health problems, which should not be used, which can be mixed together safely, and which should not be allowed to stand in a bath. This information is essential to students learning photo processes, so if they continue to do photo work in their own darkrooms at home, they will not be putting themselves and others in jeopardy.

Unfortunately, most teachers and art activity leaders know little about the chemicals they are using in photography. Labels will usually identify the chemical contents (and many include valuable warning information), but they are helpful only if the teacher already understands about the chemicals. Nevertheless, as discussed earlier, label information is extremely important in explaining as much as possible about the contents so appropriate decisions can be made. *The Compact Photo Lab Index*

---

* Orton Cones and Controllers information (www.davisart.com/safety)
† Agency location information (www.davisart.com/safety)

(Pittaro, 2nd ed. 1978) provides valuable data on precautions to be taken in photo processing. It is an extremely useful guide to the correct methods of mixing chemicals and temperature control requirements to achieve good results under safe conditions. This book is no longer in print, but can be ordered online from various booksellers.

Fundamentally, the photo teacher has the responsibility to know which chemicals should be avoided and which photo processes should not be done in any setting with children, adult students, and older adult students. The teacher or art activity leader must also be aware of the ventilation requirements for the darkroom as well as the necessary precautions for handling any chemicals.

Chemicals and Processes to Avoid

There are several chemicals that should be avoided because they are highly toxic, are difficult to handle safely, and are not essential to a quality photo experience for most students.* In most cases, these chemicals will not be found in school programs, but the photo instructor should double-check to be sure and confirm that these chemicals are not part of any regularly used compounds.

In addition to these specific chemicals, several photo processes depend on chemicals too hazardous for use in a typical non-professional program. There may be some programs with very advanced photo courses where some of the processes (e.g., color) may be appropriate. But these are exceptional instances and the teacher has a special obligation to understand all of the hazards and ways of controlling them. An excellent source of complete and highly professional information on photo hazards is *Overexposure: Health Hazards in Photography* (Shaw and Rossol, 1991). This should be used as a specific guide in solving photo program hazard problems. This book is now out of print but can be located through your local library, used bookstore, or an online bookseller.

Perhaps the best general advice is that both student and teacher always have the choice of avoiding especially hazardous chemicals or processes. For example, if a student has a thin negative, which has to be "built up" by a chemical intensifier, the choice should be to re-shoot the picture. Only rarely will the situation warrant this sort of chemical manipulation. Then, a professional photographer should do the work.

7

* Health & Safety in the Arts Searchable Data Base: Photography Techniques (www.davisart.com/safety)

Darkroom Ventilation

Professional photographers recommend general ventilation with ten air changes per hour for the darkroom. To dilute the contamination from chemicals normally in use, fresh air must be brought into the room to replace the old air ten times during every hour. Some ventilation experts feel this "rule of thumb" is inappropriate in many uses (Hemeon, 1963) and is, in any case, a difficult concept for a teacher to use meaningfully. Three steps should be taken in determining what constitutes effective ventilation:

- First, gather all possible information about the chemicals used in the darkroom, including the amount to be used at any one time.
- Second, collect information on appropriate ventilation systems and recommended standards from available sources (Shaw and Rossol, 1991; Clark, Cutter, and McGrane, 1984; ACGIH Ventilation Study Guide, 2004).*
- Third, take that information to personnel at the school responsible for designing and maintaining school ventilation systems. Get that person to apply the information to your specific room. This is not a job for the teacher alone. A conscientious teacher should know the chemicals, the number of student stations, and the size of the space. Calculating adequate air exchange and designing the system is a task for people expert in doing that.

Local exhaust configurations will probably not be necessary if the harzardous chemicals and processes can be avoided. Teachers who, for reasons of personal interest or experience, want the students to work with these processes must acknowledge the need for expensive local exhaust equipment, which, at best, will only reduce and not control the hazards. The question, of course, is whether those processes have sufficient educational value to the students to justify this cost.

Handling the Chemicals

Several simple rules should govern the handling of chemicals in photo processing:

---

* ACGIH manual source (www.davisart.com/safety)

7

- Wear gloves. Regular household gloves are not adequate protection from the chemicals, so check with chemical suppliers to get recommendations for the most effective type. Nitrile latex gloves will probably be appropriate in most cases, but do not guess.*

- Wash gloves, inside and out, after use and thoroughly wash hands as well.

- If any skin irritation should result from wearing gloves, use tongs instead, but be sure to keep hands out of the chemicals, especially developers. "Most developing baths can be highly hazardous to the skin, particularly with continued use because the effects on the skin can be cumulative" (Seeger, 1982, p. 30).

- Have a skin and eyewash facility available in the darkroom to allow immediate flushing of chemical splashes.

- Do not heat chemicals to speed their action. Heating usually causes erratic results and most often produces bad negatives or prints anyway; heating can also produce toxic chemical fumes.

- Keep all chemical baths covered when not in use so the fumes caused by evaporation are reduced as much as possible.

- Do not allow sodium thiosulphate (fixer) to become old, since it may decompose to produce sulfur dioxide. Do not heat it, and always keep it covered when not in use.

- Clean up spills so they do not dry and form dust that can contaminate the air.

In general, any process as popular and as chemically intensive as photography requires special attention to its hazards. Many students will not have had any previous art experience and some will have no subsequent classes in art, so their entire attitude about what constitutes safe practices in art will depend on what goes on in photography. It is also probable that many of those who take photography will continue to work with it in and out of unsupervised settings, so they need to be carefully instructed in the requirements for setting up a safe home darkroom.

For all of these reasons, photography teachers bear a heavy responsibility in providing complete and accurate instruction and information. The photography program should be an excellent example for all students of a safe and hazard-controlled operation.

* York University Safety Notices: Gloves (www.davisart.com/safety)

Sculpture

All three-dimensional activities are likely to be classified as sculpture. This can mean anything from a simple cut and folded paper form to a cast object involving modeling, mold making, metal heating and pouring, grinding, polishing, and adding patina. Except that the objects created in each case occupy three dimensions, and are viewed as such, they have little in common.

The sculpture processes normally done by pre-school and elementary-age children do not go much beyond various paper or cardboard constructions fastened with glue and staples, papier-mâché, simple wood constructions, clay, and perhaps wire worked in combination with scrap materials. There are of course some hazards involved with each of these, such as the use of certain adhesives, cutting tools, and the problems of any clay activities. These hazards have been discussed earlier. The sculpture activities of concern here are those likely to be done in the middle school and senior high school and by adults and older adults in non-school settings. These involve more advanced methods and materials.

Realistically, there are many sculpture processes well within the capabilities of secondary level students that are almost never done because of the cost and complexity of materials and equipment. Metal casting foundries, for example, rarely exist in high school art areas, but may in settings where classes for adults are held. Similarly, the equipment for vacuum forming plastics might be available to high school students but only in exceptional settings and those designed for adult use. Where these processes exist, the instructors must demonstrate extreme concern for the health and safety of the students. These teachers must be familiar with the highly technical information specifically directed to sculpture with those media (McCann, 1992; Siedlecki, 1972).

Similarly, using plastics to make sculpture has dropped significantly in popularity and frequency in recent years, probably in part because health hazards of the various epoxies and resins have become better known. Many of the exciting visual possibilities these materials once seemed to have, no longer are so attractive to artists, students, or teachers. Although plastics in many forms continue to have widespread commercial application, only a few professional artists, and advanced students, use them.

There is actually little to recommend plastics as a sculptural medium in any setting other than a professional sculptor's studio. Even there, it should be used only in special circumstances and done by artists knowledgeable in the health hazards. Moreover, any process involving heating or dissolving plastics for sculptural constructions should never be done at the pre-school or elementary level, or with older adults. The risks are too great, and the possibilities of controlling them too limited.

Wood Carving and Construction

These processes involve hand or power cutting tools, which have been discussed in earlier sections of this book. In addition, however, it is important to emphasize the necessity of wearing eye protection and being sure work is done in a location where non-involved students will not be injured by flying chips or errant pieces of wood.

Sanding and polishing wood are not particularly hazardous activities unless power equipment is used improperly. Belt or orbital sanders are relatively easy to control and should cause little difficulty, although large floor models sometimes move at a speed that can pull a piece of wood away from an inattentive student. All tools should be equipped with proper guards that are always used. Power sawing and sanding produce significant quantities of wood dust, much of which is suspended in the air for some period of time. Fixed sanding equipment or saws should be fitted with vacuum dust-collecting systems, and of course students should wear appropriate dust masks if there is extensive woodworking going on. Typically, there will not be enough dust in an art-making space for this to be a serious problem, but an aware teacher or art activity leader will take the proper precautions, should dusty conditions prevail.

Carving is a sculpture activity that often has great appeal to older adult students, and the precautions and concerns about the use of knives, flying chips of wood, and dust from sanding their work apply equally. Often the ways of dealing with these problems are overlooked because it is a familiar activity, but they should be stressed at every opportunity by teachers. As discussed previously, older adult students have additional health concerns that may be aggravated by or contribute to art-space hazards.

Finishing materials used with wood primarily involve solvents, which have already been discussed. Because the adhesive most often used in attaching pieces will be nontoxic white plastic glue, few problems exist in this area.

Metal Forming—Welding

In some settings, metal sculpture has grown in popularity and welding is not an uncommon practice. Although most metal work occurs in an industrial arts shop rather than the artroom, some aspects of metal forming and joining may be found there. The major hazards in working with metal involve cutting, either with handheld metal snips or large guillotine type cutting presses. Under no circumstances should the latter be used without extensive training and careful supervision. In fact, where work of this scale is undertaken, an especially thorough skill testing program should exist before students are allowed to use the equipment.

Welding is a similar case. The most common types of welding include oxygen-acetylene, arc welding, and brazing. Of these, arc welding is probably the least likely to be found in any art program, but any welding with students will require that the teacher have the most current information.[*]

Welding in any of its forms is a very specialized task and requires a fully trained and experienced teacher to prevent accidents. Welding should not be considered a process students can pursue "creatively" without a complete and thorough grounding in correct procedures. The following rules should be considered minimum guidelines for safe welding activities:

- Read and follow all warning labels.
- Always check to be sure equipment is in good working order.
- Acetylene and oxygen tanks must always be chained to a wall or to a substantial portable cart to ensure they do not fall over and damage gauges or fittings.
- Use correct protective equipment (masks, clothing, and gloves).
- Work inside only if there is an effective ventilation system and then stand so as to avoid breathing any of the fumes rising from the work. If working outside, be sure fumes blow away from other persons and building openings.
- Be very aware of the danger of fire. Keep all flammable materials well away from the work area and be sure to have an appropriate fire extinguisher within easy reach.

Welding and brazing produce various air contaminants from the rods and the metals being joined as well as from coatings or painted surfaces, which produce fumes when heated. Most of these fumes are difficult to identify, and they should all be assumed to be harmful and so require proper ventilation.

Generally, an almost unlimited number of materials and processes can be used to make sculpture. The simplest of these usually involve no unexpected hazards. However, the more complex processes, such as metal

[*] American Federation of State and County Employees: Welding Hazards
(www.davisart.com/safety)

forming and casting or work in plastics, are not ones that should be undertaken in most non-professional settings. But if they are done, the teacher must not only be skilled in the processes to teach them effectively, but be fully aware of the health and safety hazards they create for students. These problems will not be easy or inexpensive to solve, and a careful evaluation of the educational and aesthetic goals of an art program should be made to determine the importance of such activities in the program.

Jewelrymaking

Complex manipulation of materials and tools, even when doing relatively simple activities with young children, demands careful attention to potential hazards in jewelrymaking. The most important safeguard to remember is to wear goggles. Additional ways to avoid hazards in jewelrymaking:

- Be sure to file rough metal edges to reduce cuts or scratches.

- Heat metal only where local ventilation is available. In soldering or melting metals for casting, fumes should be drawn away from the student; it may be necessary to have individual elephant trunk hoods at work stations to ensure full protection.

- Wear eye protection and work only in an area with good general ventilation when polishing; be sure long hair and sleeves are securely tied back to prevent tangling in the polishing wheel.

- Handle acids used in the pickling process (nitric, sulphuric, sodium bisulfate) with great care. Wear gloves, always add acid to water in mixing the solution, and keep acid baths covered when not in use. When the acids should be disposed of, be sure to follow methods prescribed for the facility in which the work is being done—do not simply pour the diluted solution down the drain. If there is no disposal policy, request that school or program officials find out how it should be done so city water treatment facilities are not damaged. Pickling acids need local exhaust ventilation.

- Keep work areas clean and dust-free.

- Periodically check torches and all hoses to be sure they are in good condition and do not leak.

- Use only cadmium-free silver solders and fluoride-free fluxes.

- Use lead-free enamels.

- Develop rules to be followed in the use of protective equipment.

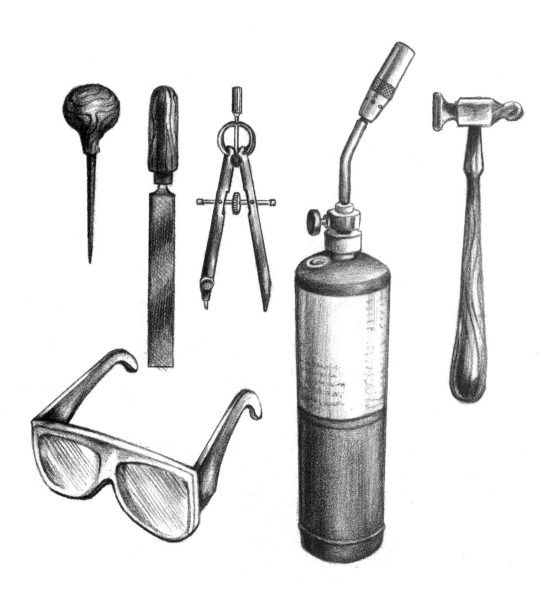

7

Perhaps the best way to introduce jewelrymaking to students is to have them design and work with very simple materials such as string, yarn, leather thongs, and found objects, which can be manipulated into body adornment. As they develop a sense of how materials can work together, wire and plastic can be explored. When they are ready to try their idea entirely in metals, they will then have a good sense about jewelry design and will find the safety requirements easier to manage. If students understand design concepts, they will have far less trouble in understanding and accepting health and safety precautions.

New Processes and Processes Not Yet in Use

There is always the expectation that some new process will be developed that can be used in art programs for children, youth, adults, or older students, and most likely it will be. By watching the work of practicing artists who will produce these future media, possibilities may become apparent. Walking through the tunnel in the United Terminal at Chicago's O'Hare airport, for example, will give an observer the thrilling sight of neon tubing used to create a gorgeous moving show of light; the hand-blown multicolored glass ceiling created by artist Dale Chihuly in the lobby of the Bellagio Casino in Las Vegas is another example of possible work. Nearly every exhibition of contemporary art will produce ideas and possibilities as well. But there is no mention of either glass or neon in this book, since both are highly technical and equipment- and space-limited processes, although the time may come when they, and other inventive and exciting media, will be possible in the settings described here.

When that time arrives, teachers and group leaders and their administrators will have an especially important responsibility they must accept. These media and ones not yet invented or used will all carry with them special hazards to both the health and safety of students. Serious research will have to be done to ensure that the processes can be made safe for non-professional artists who are working in groups, often large, and in spaces that may be severely limited, with tools and equipment that are not professional grade. It will not be until all of these issues have been addressed and the problems they create have been solved, that these, and other, future media can make the transition from artist's studio to artrooms and other limited art-making spaces.

Issues of the Health and Safety Content of the Curriculum and the Legal Problems Teachers and Leaders May Face

Chapter 8

The Curriculum and Safety in the Art Work Space

School and Community-Based Programs

The art teacher's or art activity leader's time should be spent teaching art under healthy and safe conditions that enhance student learning and provide art experiences that are rewarding and enjoyable. The role of the administration or non-school program director is to provide the available physical environment and fiscal resources to accomplish that goal efficiently. As always, it is cooperation between these parties that is essential to bring about any improvements necessary to accomplish these goals successfully.

Teachers and leaders must realize that their role is not, however, limited to working with the students in the art-making space. They must make a concerted effort to keep their administrator, whatever his or her title, informed of the problems, needs, and proposals for improving art work space health and safety conditions. In the schools of the past, art supervisors and program coordinators played a large role in planning program changes and choosing materials and equipment. Much of this work could be left to the "central office." In most schools, that centralization no longer exists, and these tasks fall upon the individual teacher and principal. While many lament the loss of a person in the district who can organize and standardize such matters, teachers have to recognize the role is now theirs and accept the responsibility for it.

The concerns for those who teach in pre-schools are much the same as for other schools, since pre-schools are, for the most part, licensed and have regulations that cover much of what they do. Health and safety rules may not, however, cover concerns about art materials and so the teacher must make sure that the administrator understands the importance of being aware of potential problems.

For an art activity leader in any setting, working with a few children on a limited basis, it is easy to overlook problems that might exist. Generally, there is no supervisor as such and the responsibility is totally on his or her shoulders. This makes the selection of art-making activities a critical task that must be taken seriously and generally there is no shared responsibility. In particular, the selection of projects should be governed by the availability of entirely nontoxic materials.

People who work as teachers of adult students, perhaps in classes in their own studios, must also assume that same responsibility on their own. Being conscious of the hazards of the processes, teachers must be sure to teach that information to students. These teachers should discuss with their students any health problems, allergies, or sensitivities they may have to be sure nothing they do will exacerbate those problems. Teachers should do this at the beginning of each class session so that students will not buy or keep materials that are hazardous. Overlooking this step is doing a serious disservice to the students.

For those who provide art and craft experiences for older adults, usually in senior centers, there is a further obligation. The students in these classes are in a high-risk group; that is, they may bring to the class such problems as heart defects, pulmonary illnesses, or other physical conditions such as Parkinson's, poor eyesight, hearing loss, and mobility limitations. Each of these presents special concerns for the art activity leader and must be given very careful and thorough consideration. Confer with the program director and try to determine the kinds of hazards that need to be addressed: good ventilation, excellent lighting, free and easy access, space to work, and space to store materials and projects they cannot easily transport themselves. Kiln location and ventilation are critical; unvented kilns can produce fumes and gases that can permeate not only the art work space itself, but the entire facility. This responsibility cannot be taken lightly, and the teacher's need to know is critical to create a program that does more good for these students than it does harm.*

No one can tell a teacher or art group leader what type of health and safety advocacy will work best with a specific principal or program director.

---

* Kiln ventilation information from AMACO (www.davisart.com/safety)

Administrative style and personality vary greatly, and the leader's ability to "read" that style and personality is fundamental to any effort to gain support. There are some basic actions, however, that may be effective in achieving results:

- Organize a health and safety program including input from student groups, procedures for the inventory and investigation of all art materials, specific housekeeping practices to reduce hazards, a procedure for testing every student's understanding of the issues and keeping a complete record file.

- Assess the art program and art projects to identify hazardous activities and either develop plans to cope with the hazards or replace the activity with one of equal value.

- Keep the program administrators (if there are administrators) fully informed of plans, actions, and accomplishments. Be sure that these administrators are aware of and knowledgeable about policies and regulations that may be applicable to your teaching situation. These may involve OSHA, NIOSH, and EPA guidelines and state department of education policies, or may involve fire regulations applicable to the setting in which your classes are being conducted.

- Teachers, art activity leaders, and program administrators need to know the requirements for reporting incidents of accident or illness to the appropriate administrative agencies. A clear policy should be prepared that is available to keep all involved persons informed so there is no question as to the correct procedures.

- Request expert advice on technical matters. Don't call in the fire marshal without the administrator's knowledge, but be aware that specialized answers can come only from specialists.

- Involve other teachers and school and community personnel in the health and safety program.

- Seek advice from the school nurse or medical personnel who can provide information and guidance related to the ages of the students in the group with which you are working (e.g., pre-k or the elderly).

Effective health and safety program work will benefit from involving people outside the art program or activities. It gives them the opportunity to share special knowledge and skills with the students and to feel they are making an important contribution to the schools and other community

8

programs. It is also possible that this involvement will enlist support not only of a health and safety effort but of the role of art in the school curriculum and in the creative lives of all of a community's citizens.

Some art programs result from popularity. Unless certain courses are offered, enrollment will drop, and the art class will be eliminated or classes cancelled. Certainly, this is especially true of non-school programs where students are in classes completely by choice.

Some art programs aren't really programs at all, but the product of expedience. They comprise projects that caught the teacher's fancy, were interesting to the students, or satisfied the needs of some holiday. This is often the way in which projects are generated in non-school settings and after-school art activity settings. Often these projects only serve to keep students busy and may have nothing to do with one another and little to do with learning about art. And some art programs are a response to pressure—the expectations of the parents, the program administrator, or a curriculum guide or program plan.

Most art programs are a mixture of all the above, and, if the mix is a sensitive one, they may be very good. There is certainly nothing wrong with meeting student interests or parent expectations, but teachers must plan programs carefully. If there are health or safety hazards involved, they must weigh these problems against the benefits of meeting expectations that may be of dubious value. The question always to ask is if the activity is essential to the goals of both the program and the school.

Careful program planning is thus critically important in determining what hazards must be surmounted. Good planning also makes it infinitely easier to justify any special costs in doing so. A teacher or art activity leader who rationally demonstrates that an activity is essential will be effective in rallying support for safety improvements.

Proposing Changes: Strategy and School and Community Action

As has been emphasized in earlier sections of this book, do the easy things first. This will initiate a strategy to bring about a healthy and safe art-working-space environment. For example, if ventilation is thought to be the most serious problem in a particular artroom, demanding that a ventilation system be installed immediately may not be the best tactic. First, remedy the secondary factors that contribute to the ventilation problem. Because ventilation systems are usually expensive, such requests will hold more weight if all other possible actions have been taken to reduce the problem of poor air quality; this is what is meant by secondary factors.

Demonstrating a commitment to alleviating the problem on all fronts will help convince the administration or building owner that a new ventilation system is a necessity. But remember, no ventilation system can blow away problems that originate in program planning or instructor practices that are indifferent to the creation of hazards.

Each type of program has a somewhat different method for instituting changes in such things as assigning rooms, allocating space for storage, and setting capital improvement priorities. Art teachers and art activity leaders must work within the normal channels to garner support for their requests. An artroom may need more storage, a separate room for clay mixing, local ventilation for kilns, or significant replacement or addition of tools and equipment. Such projects always compete for dollars with such other proposals as improvements to the cafeteria, installation of handicapped facilities, repair of furnaces, replacement of broken windows or a gymnasium floor, and even the building landscaping. It is formidable competition. To compete successfully, consider the following:

- Learn the system. Find out what things influence decision-making. Is it cost, enrollment figures, pressure from outside groups, or a bias toward some areas as opposed to others? Without understanding both the system and the politics behind it, proposals will have only a slim chance of being approved.

- What are upward limits for any single project? Can projects be split into phases and receive a funding commitment for a two- or three-year period?

- What external support would be helpful? Would it harm the proposal's chances to seek PTA or community groups to aid in either advocating or fund raising?

- Who are the people most influential in making the decision, and how can they be convinced of the proposal's importance? If the program is city-funded, who makes the actual budget decisions and where is the most effective place to focus lobbying efforts?

There is no sure way to succeed in funding a health and safety proposal. Even in the best of economic circumstances, correcting hazards in an art work space is never a sure thing. There is, however, no substitute for a meticulously prepared plan: the successful teachers and art activity leaders will have done their homework thoroughly. Art classes usually don't have the luxury of the almost automatic support often given reading, science, or

8

athletic activities, nor support that comes from student testing results, so the task will not be easy. Two additional points should be made concerning strategy: first, do not use scare tactics, and second, don't provide the solutions for the problems.

If the problems are presented in a strident and threatening manner, any art activity program will very likely suffer. Claiming that the problems can only be solved with large quantities of money may only convince program administrators it isn't worth the cost.

Providing solutions to complex problems should be left to experts; teachers and art activity leaders should only compile the data about their students' use of problem materials and about the activities that take place, but leave solutions to experts. Some teachers may be very well informed about hazards, but, nonetheless, do not have the technical knowledge to correct them. Moreover, should an instructor-designed system go wrong, there usually is no possibility for further corrections. Recognizing the problem and designing the solution are different jobs, and the wise art activity leader will not confuse them.

The Value in Substituting Materials and Procedures

The substitution of nontoxic for toxic materials has already been emphasized throughout this book as an important way to control hazards. What has not been said, however, is that substitution of materials is not a step that can usually be taken quickly. The substitution may involve changing activities or changing the kind of materials used within an activity. In any case, the change-over requires the selection and ordering of safer, clearly labeled materials that will take time to be ordered and delivered. Unless the decision to change materials is made at exactly the right time in the budgetary cycle, it may be as much as a year or longer before the substitute supplies arrive for use. Fortunately most materials used in art making are now readily available in nontoxic form and generally are no more expensive than their older, less safe counterparts, so ordering is easier, but timing orders is still important.

This is not intended to discourage substitution practices, but rather to emphasize the need to plan ahead. How can the changes be made as expeditiously as possible? The key is the person responsible for ordering supplies for the program. It may be necessary to discuss procedure with the school principal or the non-school program administrator, but usually the district purchasing agent (who in some cases may be that school principal or program administrator) will cooperate to ensure that the new materials are safer than the old ones and explain how specifications must be written in order to receive what is expected.

It is essential that if nontoxic materials are not available for the activity that is planned, and the decision to continue using unlabeled materials is made, the person doing the purchasing should help in getting the information you need. Ask him or her to get a Material Safety Data Sheet (MSDS) from online sources to get toxicity information in the event any specific material is not labeled.* This will not occur often since all art materials made since 1986 must at least carry an ASTM 4236 indicia indicating the material conforms to the nontoxic labeling standards of the American Society for Testing of Materials.†

The MSDS describes the potential hazards of materials and provides other information related to dealing with their effects. In the event the MSDS is not available online, art materials manufacturers have both a moral and legal responsibility to their customers to provide one. Having the information on these forms is the only way to know exactly the degree and nature of hazards that materials present to students. But in all circumstances and all settings, use materials only when you are familiar with their contents.

Last-Resort Solutions

Sometimes all efforts fail to correct a hazard: the substitution of activities or materials may not be desirable; cleaning the artroom does not accomplish the expected effects; there are no funds to make physical improvements or to remodel the existing room; perhaps the classes are just too large for the space. Every teacher must face the possibility that sometimes the hazards cannot be eliminated by these means.

In this situation, there are still two possible courses of action: either totally change the program, or change either the location of the artroom or the location of the hazardous activities. In all probability, moving will be the best choice.

Getting support to move may not be difficult if the case is well made. The teacher must of course document the need, suggest new locations, and outline the benefits and costs of such a move. However, the purpose of such a move must not be to increase space or improve conditions other than those related to solving a serious health or safety problem.

8

---

* MSDS online source (www.davisart.com/safety)
† ASTM statement (www.davisart.com/safety)

Chapter 9

Legal Liability Issues and the Legal Implications of Art Work Space Hazards

Genuine concern for students should be sufficient motivation for teachers and group leaders in any art-making setting to make health and safety an integral part of their art programs. But the possibility of becoming a defendant in a lawsuit may be needed to move some teachers to recognize their obligations to students whether they are in a school or non-school setting and whatever the age of their students may be. In a society that has increasingly turned to the courts to solve problems, lawsuits are an ever-present reality for teachers. Judgments in the hundreds of thousands of dollars are common in damage actions, and the number of times teachers are defendants in such cases can be expected to increase in the future.

Art teachers in school artrooms, leaders of after-school and evening groups of children who may be making art, and teachers of adult and older adult students all need to be aware of their vulnerability to legal action and of steps they should take to protect themselves. While it doesn't apply to adult and older adult students, those working with children and young people must be aware that "Because of the age, maturity, and experience limitations of most students, and because teachers stand *in loco parentis* (in the place of a parent), schools are expected to provide a safe environment and establish reasonable behavior for the protection of students" (Florio, Alles, and Stafford, 1979, p. 222).

Teachers working with adults are not unaffected by problems of liability and thus should pay close attention to the factors that may cause it to become an issue for them. "Torts are defined in the law as civil wrongs 'for which a court will afford a remedy to the injured party in the form of damages'" (Alexander, 1984, as quoted in Sewall, 1995). Negligence is the tort most likely to be brought against an art teacher or art activity leader, whatever the setting. In cases where students were injured in the artroom, the teacher has often been sued for negligence on the grounds that the injury wouldn't have occurred had there been proper instruction and supervision: the teacher should therefore be liable for damages.

In general, there are several defenses a teacher can use in court. But no examples in school law literature specifically refer to art teachers or to injuries occurring in artrooms or other art-making spaces. Moreover, cases in which the damages were sought for negligence resulting in illness from exposure to toxic materials, rather than injury, in the art-making setting are likewise not mentioned. There is little comfort in this for anyone teaching art. Everyone should take extra precautions because even though precedents do not seem to exist, there may very well be someone seeking to establish them.

However, "it must be emphasized that a teacher is not liable for personal damages every time a student under the teacher's supervision is injured in an accident" (Flygare, 1976, p. 42). It is important to know under what circumstances the teacher can be held liable. An affirmative answer to each of the following four questions indicates liability:

1. Did the defendant (art teacher or art activity leader) have a duty or responsibility to the plaintiff? Was there an obligation to protect the plaintiff against unreasonable risk of harm?
2. Was there a failure to conform to the required standard of care owed the plaintiff? This question establishes the existence of negligence, which is the dominant principle of tort law.
3. Was there a causal connection between failure to provide adequate care and the resulting injury (proximate cause)?
4. Did an actual loss or damage result from the commission or omission of an action by the defendant? (Prosser, 1971, quoted in Florio, Alles, and Stafford, 1979, p. 220)

Teacher or art activity leader action:* Persons teaching or leading art-making activities need to relate each of these four questions to the art-

---

* Following each of the descriptions of legal responsibility an art teacher or art activity leader has in working with students, is a statement describing what can be done to reduce the possibility of error and accident: "Teacher or art activity leader action." This is not legal advice, but a commonsense interpretation of what teachers and leaders can do to ensure no legal issue is likely to be raised against them and also to be sure students' health and safety is not placed in jeopardy.

space situation and take steps to be sure the answers are negative. This is the best way to protect against tort action.

Duty to Protect or Teacher Responsibility

Another way to describe and identify the teacher or art activity leader's duty or responsibility is that the injured party must provide answers to those same questions framed in this way to establish negligence:

1. That the defendant (art teacher) had a duty to protect the injured person.
2. That the defendant failed to do that.
3. That the failure was the proximate cause (a substantial factor) of the injury (Peterson, 1978, p. 252).

Duty involves the issue of "standard of care," and it has been interpreted that teachers "are held to be a higher standard of care than the ordinary man on the street. The teacher or (program) administrator is under the duty to possess more than the 'ordinary' amount of intelligence in relation to students and their care" (Gatti and Gatti, 1975, p. 213). Thus, "teachers have a duty to anticipate foreseeable dangers and take necessary precautions to protect students in their care" (McCarthy and Cambron-McCabe, 1992, quoted in Yell, 1999). "School districts should take actions to make certain that administrators, special and general education teachers, and other personnel are aware of their care and supervisory duties under the law" (Daggett, 1995; Freedman, 1995; Madsley, 1993, as quoted in Yell, Mitchell L., "What Are Tort Laws," 1999).

The required standard of care may be considered less for high school students or for adults than for elementary or pre-school students because of differences in age and maturity. "It is not uncommon for courts to consider that children under the age of seven do not have the ability to make appropriate decisions for themselves and that even in the circumstances in which a teacher has painstakingly given a student instructions as to safe behavior, the charge of negligence may not be precluded should the student be injured. Similarly, an assumption on the part of a teacher that a child age 7 through 14 understands all instructions concerning safety may be invalid, as the level of understanding and ability to make good judgments possessed by students in this age range usually must be assessed on an individual basis. Students over the age of 14 are generally assumed by the courts to be able to understand instructions given them absent some

9

disability which might impair the individual student's judgment" (Sewall, 1995).

Teacher or art activity leader action: Being aware of the differences in students is critical to understanding the standard of care that is applicable. Knowing, for example, that pre-k children have not fully developed intellectually or physically places a much higher standard on teachers or art activity leaders at this level. Working with the elderly, who may have physical infirmities related to age, also raises the standard of responsibility.

Proximate Cause

Proximate cause means that "there was a connection between the breach of duty by the teacher and the student's injury" (Yell, 1999). For example, if the teacher omitted some instruction, and the injury would not have occurred had the instruction been given, then it was "proximately caused" by that failure and there has been negligence on the part of the teacher.

Teacher or art activity leader action: It becomes especially important, therefore, for the teacher not to assume the students are familiar with any activity or tool use. Always give instructions pertaining to them. The question that will be asked in court is, "Did the injury occur because of something the teacher did or did not do?" For a teacher to be held blameless, it cannot be found that something was overlooked when the students were given instructions for their work.

Foreseeability and Teacher-Made Rules

"When a teacher or administrator foresees or reasonably could foresee that an injury might occur if a particular condition is not corrected or preventative action is not taken, he or she has a duty to do something about it before the injury occurring. If he or she does not, liability may be imposed for negligence" (Gatti and Gatti, 1975, p. 133).

Teacher or art activity leader action: The "reasonable and prudent" teacher must ensure that the problems are corrected before injury or illness occurs and establish rules about student conduct where a problem situation exists. If the condition cannot be corrected by the teacher, a formal written request to have it done should be made to the school or program administrator immediately with a copy kept for the teacher's own records.

Teachers have the right, as well as responsibility, to make rules enforcing safe procedures. To be legally binding, the rules must be written, should be clearly understandable and concise, must be communicated to the students, and must be enforced. "Many teachers are sued and held liable for simply not having enforced rules regarding health and safety of their students. . . . Remember, the rules must be reasonable, lawful, and not conflict with the student's constitutional rights" (Gatti and Gatti, 1975, p. 226).

Contributory Negligence and Assumption of Risk

If a student is injured by behaving in a manner that jeopardizes his or her own safety, that behavior may be defined as contributory negligence. If a student is guilty of contributory negligence, no damages can be recovered. But the burden of proving that negligence is on the defendant (in this case, the teacher), who must show that the student's behavior in part caused the injury. In this type of defense, it must be determined whether the student, given his or her age, could be expected to understand the nature and extent of the danger. If the student could not be expected to know, then the behavior would not be considered contributory.

Assumption of risk is a legal defense based on the contention that the student assumed the risk of injury by participating in the activity. But "an essential requisite to invoking the assumption of risk doctrine is that there be not only knowledge of a physical defect in the premises, but also appreciation of the danger produced by the physical defect" (Peterson, 1978, p. 256).

Teacher or art activity leader action: In respect to the art-making environment, this defense depends heavily on whether the student was properly instructed about the hazards involved in the activity and whether the student actually had that knowledge.

Intervening Acts

If a teacher actually has not fulfilled his or her responsibilities in exercising the proper care, damages would still probably not be assessed in a negligence case if some intervening act was found to be the proximate cause of the injury. "In some instances an intervening event, such as the negligence of a third party, has relieved school personnel of liability" (McCarthy, 1981, p. 176). However, a judgment might still be rendered against the teacher in such a case if there is reason to believe the teacher should have anticipated and prevented that intervening act.

Teacher or art activity leader action: Suppose, for example, a teacher has failed to provide careful instructions and warned about the dangers in using a paper cutter. In an art-making incident, a student loses a finger

9

directly as the result of careless behavior of another student at the paper cutter. Although the second student's action would be an intervening act, there may be reason to believe the teacher is still liable if it can be shown that teachers should always anticipate problems when two students use the cutter at one time. Obviously, hoping there may be an intervening act should hardly be an excuse for inadequate initial instruction, and teachers should always give appropriate instruction and carefully monitor the behavior of the students at all times.

Waiver of Liability—Permission Slips

It is common belief that to have a parent sign a permission slip for a student activity constitutes a waiver of the teacher's liability in the event of student injury or illness during that activity. However, "negligence is not among the risks accepted by parents on behalf of their children" (Florio, Alles, and Stafford, 1979, p. 224), and a permission slip simply indicates the parents are aware the child is participating in the activity. The principal value of such a slip, if an injury does occur and there has been no teacher negligence, is that it can be used as a defense against any charges by the parents that the activity was not one in which the child should have participated. If, however, negligence can be proved, the teacher is still liable for damages despite the existence of a signed permission slip.

Teacher or art activity leader action: Teachers and art activity leaders should consider that permission slips are actually little more than a way of notifying parents or guardians that some excursion is going to take place and what it will entail. It would be a good idea to include a place for the parents or guardians to acknowledge that they understand the risks and that they are aware an excursion is being made.

Providing Medical Assistance

Giving medical assistance may result in litigation if it is not deemed reasonable and prudent under the circumstances. In an emergency, the teacher or art activity leader is expected to provide first aid if no medical personnel are available or if the injury is such that attention cannot wait for the arrival of professional medical assistance. But a person acting in an emergency "cannot be held to the same standard of care as one who has had time to reflect. Even if it later appears that the defendant made a decision which no reasonable person could possibly have made after careful deliberation, there is no negligence" (Prosser, quoted in Florio, Alles, and

Stafford, 1979, p. 223). Not to give any treatment in an emergency situation, however, is considered negligent, but any treatment must not cause the condition to become worse. If that should happen, the teacher might be held liable for causing a more severe injury than was originally incurred.

Teacher or art activity leader action: It is important to have at least first aid training so that the immediate response is not going to do more harm than good. The apparent seriousness of the injury or illness will be a major factor in response, but it is always better to err on the side of caution and to notify the parents immediately. Obviously, if the injury is major, seek medical help at the first opportunity, provide first aid, and then notify the parents.

Avoiding Litigation

There are several actions teachers can take to reduce the chance of a lawsuit charging negligence. The best, of course, is "the prevention of injury through competent instruction and adequate supervision" (Florio, Alles, and Stafford, 1979, p. 25). However, several other actions are of equal importance:

Teacher or art activity leader action: Ensure that dangerous conditions have been eliminated.

1. Maintain the room in a safe manner with equipment properly located for safe use and be sure that hazards arising from the room itself are identified and eliminated.
2. Keep all equipment in good working order: sharp cutting edges, electrical cords in good condition, proper kinds of protective gear available.
3. Store materials and tools in appropriate containers and/or cabinets and clearly identify the contents.

Teacher or art activity leader action: Establish rules of behavior and enforce them.

1. Make sure rules are clear, concise, and understood.
2. Put rules in writing and either distribute them to the students or, if appropriate, post them in the area where the activities they govern take place.
3. Require responsible behavior. Do not allow immature actions to go unchecked, and be sure all offenders are disciplined.

9

Teacher or art activity leader action: Formally test students' understanding of correct procedures.

1. Be sure students are carefully instructed in the correct use of all materials and equipment.
2. Develop and use tests to verify students have both knowledge and skills to participate in art activities. (See Figures 14 and 15 for test examples.)
3. Prohibit participation in activities unless the tests have been satisfactorily passed.

Teacher or art activity leader action: Maintain complete records of the health and safety activities that are part of the regular art program.

1. Keep copies of all inventory lists, condition reports, requests for elimination of room hazards, student tests, permission slips, and information on allergies.
2. Keep lesson plan copies indicating that health and safety instruction was planned and actually provided in each class session.
3. Maintain a complete description of the artroom health and safety program, plans for using warning and information signs, and copies of the constitution of student health and safety committees and minutes of any meetings.

What About Liability Insurance?

There is little question that even if a teacher meticulously follows all these suggestions for avoiding litigation, there remains the possibility that someone may file a lawsuit charging negligence. Because the laws in a number of states provide protection from litigation against school districts (known as the principle of sovereign immunity), it may actually be illegal for the district to provide liability insurance for its employees. The reasoning is that since the district can't be held liable, it cannot spend district funds for liability insurance.

Teacher or art activity leader action: Each teacher should make an effort to find out the liability coverage that may be provided as a part of employment. If there is none, seek other sources for it. Most professional organizations make liability insurance available to members at reasonable cost (for example, the National Art Education Association [NAEA] has such insurance available to its members). Other sources may be available

Sample Competency Essay Test

COMPETENCY TEST: **Electric Kiln Operation**

Name: _____

Date: _____ Class: _____

Teacher: _____

---

1. Describe each step that must be taken to safely bring the kiln to full firing temperature once it has been loaded. Be very precise in your answer.

   a. _____

   b. _____

   c. _____

   d. _____

2. How long must a kiln cool before the lid may safely be opened?

3. What kind and amount of room ventilation is required when an electric kiln is being fired?

4. What specific precautions must be taken in checking cones during kiln firing?

5. (Optional, if appropriate.) Describe the steps that must be taken to activate the automatic electric kiln control equipment.

This example should be modified to fit specific settings, expected student responsibilities, and the type of equipment used.

Figure 14

9

Sample Competency Test—Multiple Choice and Observation

COMPETENCY TEST: Linoleum Printmaking Tools

Name: _____

Date: _____ Class: _____

Teacher: _____

Teacher Observation of Student Skills (Initial Approval) _____

Check as many answers as you believe are correct.

1. Linoleum blocks are only to be cut using what tool?
   a. n Paring knife   c. n Linoleum cutting tools
   b. n Linoleum knives   d. n Razor blade

2. A bench hook is used for what purpose in linoleum block?
   a. n To hang coats on benches
   b. n To support the block while cutting
   c. n To print the block
   d. n To hook the linoleum to the bench

3. How can you tell when a cutting blade is too dull to use safely?
   a. n It slips across the block without cutting
   b. n It doesn't cut the block easily
   c. n It takes on a dull color
   d. n It has been discarded by someone else

4. If it is necessary to hold the linoleum block with your hand, how should it be held?
   a. n In front of the cutting tool
   b. n Between the cutting tool and your body
   c. n Away from the path of the cutting tool
   d. n It doesn't need to be held

5. What part(s) of the printing press can most likely cause injury when the press is in use?
   a. n The turning spokes   c. n The bed
   b. n The rollers   d. n The press table

Note to student: When you have finished this part of the test, notify your teacher and arrange to demonstrate the use of the linoleum block.

Multiple choice questions are quick to evaluate and can be used effectively because there are specific answers. When student judgment is required, short-answer essay tests are more effective. Combining multiple choice with observation increases the effectiveness of the test.

Figure 15

to part-time employees of a city or community college program, but seldom is there a group policy available for the individual teaching in his or her own studio. It is important to check if a homeowner or rental policy can provide the coverage you need. Determine if the coverage pays legal fees as well as any judgments that might be rendered. Do not be shy about seeking substantial coverage: judgments in tort actions often reach the hundreds of thousands of dollars, and legal fees can be very heavy.

A good teacher, careful in health and safety instruction, will take this extra step to ensure against ruin. Liability insurance will not ease the emotional pain a teacher feels from knowing a student may have been handicapped for life as a result of an art-making space accident. But the lack of adequate insurance may make a tragic situation even worse.

Controlling Record Keeping

No teacher wants to add to an already heavy record-keeping burden, and art teachers seem to be in a special class when it comes to finding reasons to avoid paperwork. There is no question that filling out forms and maintaining files strike most teachers as busy work and an eminently avoidable waste of time. This attitude should not prevail against health and safety program records.

Maintaining a safe workplace to protect oneself from litigation ought to be reason enough for teachers to keep the required records. However, there are ways to keep this task reasonable. Once the initial information has been collected, only a small amount of time will be required to keep everything current. Use student help in gathering that initial information and set aside a specific time each week (or two) to update your information. In this way you can control the records you keep and they will not control you.

Use a computer for as much data storage as possible. Download to your computer the several forms in this book, which will help you achieve this control. Also consider keeping as much individual student information as possible on a computer in the same way you may already be keeping grade and attendance records.

Teacher or art activity leader action: Assign one file drawer strictly for hard copies of health and safety material. Divide the drawer into three separate sections:

1. One should contain all material relating to the overall health and safety program, such as a general plan, the organization of health and

9

safety committees, and statements about reporting procedures for health and safety activities.

2. One should be used for folders holding blank forms to make inventories and condition checks, tests, requests for information, permission slips to be signed, and reports to be forwarded to the school administration.

3. One should hold individual files containing completed forms in all the categories mentioned above. There should also be folders for copies of any reports that have been submitted to anyone about any aspect of the health and safety program.

Teacher or art activity leader action: Use student help to make inventories and condition checks. However, be sure to verify their observations and initial their reports to indicate agreement with their judgment.

Teacher or art activity leader action: Regularly and habitually include comments on health and safety instruction in lesson plans; keep those plans indefinitely.

Teacher or art activity leader action: When classes have been completed at the end of each semester or year, take all of the folders with records pertaining to the students, tie them into a bundle, label them clearly, and store them in the school archives. If the school does not have such a place, take them home and store them where they can be found easily. There would be little reason to keep such records if the only concern was injury from accidents in the art-making space because those accidents would already be known and dealt with. But, with the increasing awareness of chronic illness resulting from exposures to chemicals in art processes, it would be well to maintain these records for a much longer period of time.

Being the defendant in negligence litigation is a very real possibility for any art teacher or art activity leader. Even if the record keeping is tiresome and time-consuming, it should not be considered too much trouble. In this regard, there are two pieces of advice that every art teacher should heed:

Be a careful teacher or art activity leader who is aware of health and safety hazards in the art-making space, who instructs students fully, tests skills and knowledge adequately, and monitors behavior constantly.

Be able to prove it.

Bibliography

Barazani, Gail C. *Safe Practices in the Arts and Crafts: A Studio Guide.* New York: College Art Association of America, 1978.

Burton, D. Jeff. *Companion Study Guide to Industrial Ventilation: A Manual of Recommended Practices.* American Conference of Governmental Hygienists (ACGH), 2004.

Carnow, Bertram. "Natural Disease—Unnatural Cause," address. Health Risks in the Arts, Crafts, and Trades Conference: Chicago, April 2, 1981.

Clark, Nancy, Thomas Cutter, and Jean-Anne McGrane. *Ventilation: A Practical Guide.* New York: Center for Occupational Hazards, 1984.

Doull, John, Curtis D. Klassen, and Mary O. Amdur, eds. *Toxicology.* New York: Macmillan Publishing, 1980.

Florio, A. E., W. F. Alles, and G. T. Stafford. *Safety Education.* New York: McGraw-Hill, 1979.

Flygare, Thomas. *The Legal Rights of Teachers.* Bloomington, IN: The Phi Delta Kappa Foundation, 1976.

Gatti, R. D., and D. J. Gatti. *Encyclopedic Dictionary of School Law.* West Nyack, NY: Parker Publishing, 1975.

Hemeon, W. C. L. *Plant and Process Ventilation.* New York: The Industrial Press, 1963.

Loeffler, J. J., ed. *Industrial Ventilation,* 18th Ed. Lansing, MI: American Conference of Governmental Industrial Hygienists, 1984.

McCann, Michael. *Artist Beware.* New York: Lyons and Burford Publishers, 1992.

McCarthy, Martha M., and Nelda H. Cambron. *Public School Law.* Boston: Allyn and Bacon, 1981.

McElroy, Frank E., ed. *Accident Prevention Manual for Industrial Operations.* Chicago: National Safety Council, 1969.

NIOSH. Preventing Health Hazards from Exposure to Benzidine Congener Dyes. Washington, D.C.: U.S. Department of Health and Human Services, 1983.

———. *Welding Safety.* Washington, D.C.: Government Printing Office, 1980.

———. *A Painter's Guide to the Safe Use of Materials.* Chicago: The Art Institute of Chicago, 1982.

———. *A Photographer's Guide to the Safe Use of Materials.* Chicago: The Art Institute of Chicago, 1983.

Peterson, L. J., R. A. Rossmiller, and M. M. Voltz. *The Law and Public School Operation,* 2nd Ed. New York: Harper and Row, 1978.

Pittaro, Ernest M., ed. *The Compact Photo Lab Index,* 2nd Ed. Dobbs Ferry, NY: Morgan and Morgan, 1978.

Seeger, Nancy. *A Ceramist's Guide to the Safe Use of Materials.* Chicago: The Art Institute of Chicago, 1982.

Sewall, Angela Maynard. Monograph. "Teacher Liability: What We Don't Know Might Hurt Us." Dept. of Educational Leadership, University of Arkansas at Little Rock, 1995.

Shaw, Susan, and Monona Rossol. *Overexposure: Health Hazards in Photography.* 2nd Ed. New York: Watson-Guptill Publications, 1991.

Siedlecki, Jerome T. "Potential Hazards of Plastics Used in Sculpture." *Journal of the National Art Education Association.* Vol. 25, February 1972.

Spandorfer, Merle, Deborah Curtis, and Jack Snyder. *Making Art Safely.* New York: John Wiley and Sons, 1996.

Yell, Mitchell L. "What Are Tort Laws?" Center for Effective Collaboration and Practice, 1999.

Index